82 6600510 8
TELEPEN
AF765616
WITHDRAWN FROM STOCK
CPC UCW
LLYFRGELL
LIBRARY
ABERYSTWYTH

COMPUTERS IN LIFE SCIENCE RESEARCH

FASEB MONOGRAPHS

General Editor: KARL F. HEUMANN

Volume 1 • THE SCIENCE OF LIFE: Contributions of Biology to Human Welfare
Edited by K. D. Fisher and A. U. Nixon

Volume 2 • COMPUTERS IN LIFE SCIENCE RESEARCH
Edited by William Siler and Donald A. B. Lindberg

Volume 3 • BIOLOGY OF AGING AND DEVELOPMENT
Edited by G. Jeanette Thorbecke

A Continuation Order Plan is available for this series. A continuation order will bring delivery of each new volume immediately upon publication. Volumes are billed only upon actual shipment. For further information please contact the publisher.

COMPUTERS IN LIFE SCIENCE RESEARCH

Edited by
William Siler
University of Alabama in Birmingham

and
Donald A. B. Lindberg
University of Missouri

FASEB, Bethesda
PLENUM PRESS, New York and London

Library of Congress Cataloging in Publication Data

Main entry under title:

Computers in life science research.

(FASEB monographs; v. 2)
Proceedings of a conference sponsored by the Federation of American Societies for Experimental Biology, the National Science Foundation, and the National Institutes of Health.
"The material in this book originally appeared in Federation proceedings, volume 33, no. 12, December 1974."
Includes bibliographies and index.
1. Life sciences—Data processing—Congresses. I. Siler, William, 1920-
II. Lindberg, Donald A. B. III. Federation of American Societies for Experimental Biology. IV. United States. National Science Foundation. V. United States. National Institutes of Health. VI. Series: Federation of American Societies for Experimental Biology. FASEB monographs; v. 2. [DNLM: 1. Biomedical engineering—Congresses. 2. Computers—Congresses. WI F202 v. 2/QT34 C738]
QH324.2.C65 1975 574'.028'54 75-34075
ISBN 0-306-34502-1

COLEG PRIFYSGOL CYMRU
THE UNIVERSITY COLLEGE OF WALES
ABERYSTWYTH
ACC. No. B78-3116
CLASS No. QH55 C7

The material in this book originally appeared in *Federation Proceedings* Volume 33, No. 12, December 1974. First published in the present form by Plenum Publishing Corporation in 1975.

Copyright © 1974, by the Federation of American Societies for Experimental Biology

Plenum Press, New York is a Division of Plenum Publishing Corporation
227 West 17th Street, New York, N.Y. 10011

United Kingdom edition published by Plenum Press, London
A Division of Plenum Publishing Company, Ltd.
Davis House (4th Floor), 8 Scrubs, Harlesden, London, NW10 6SE, England

All rights reserved

No part of this book may be reproduced, stored in a retrieval system, or transmitted, in any form or by any means, electronic, mechanical, photocopying, microfilming, recording, or otherwise, without written permission

Printed in the United States of America

Contents

Session III
Computer equipment configurations

Session IV
Data base systems

Session V
Networks and shared facilities

Session VI
Meeting the costs

Keynote Address

DONALD A. B. LINDBERG, *Conference Chairman*
University of Missouri, Columbia, Missouri 65201

Welcome to Aspen and to the FASEB Conference on the computer as a research tool in the life sciences. I want personally to thank the members of the Steering Committee, the Program Committee, and the FASEB Staff for their diligent and constructive work toward the aim of a successful meeting, and the funding agencies for their encouragement and support. We are also indebted to Dr. Donald King of the Given Institute for allowing us to meet here. The real success of such a meeting will naturally be attributable to you—the speakers and the participants.

The formal program of papers and the names of all concerned are before you already. Please note that we have tried to provide ample time and opportunity for you to meet and consult with whomever you wish during the meeting. Since this group includes experts in life science research as well as experts in more purely computational techniques, a profitable exchange of views and strategies does not seem at all an unreasonable expectation. At the end of these remarks I will mention some specific opportunities for informal consulting during the conference.

Beyond the level of personal intellectual gain, there is the question what as a group we might hope to attain at this conference. We do have some real expectations for the group's accomplishments.

First, the meeting ought to result in a formal statement which assays the past and present contributions of computing to life science research. This will of necessity be limited to a brief sampling of what is now quite a broad research area. Even so you will want to listen closely to the presentations and to note areas in which the efficiency of the investigator was increased by using the computer, areas in which the creative output of scores of researchers was brought together, preserved, and even synthesized through the use of a common computational technique. You will also want to be able to identify significant biological and chemical phenomena that could not have been observed at all without the computer as a tool.

Second, we hope that through the efforts of this group of presentations and participants that you can identify some high-leverage areas. That is, to recognize research and computational areas in which the groundwork for a major contribution has already been completed—and in which success is dependent on consistently applying known efforts and techniques. A substantial set of presentations will describe work, the costly foundations of which have been paid for, and in which ultimate success depends largely on simple survival of existing research groups.

"How to survive" is in an oversimplified sense our third expectation. That is, we hope jointly to exchange ideas on how we may obtain access to computing services—or in some cases, continue to get computing accomplished in the face of what are clearly perilous times for any research group. To many researchers in the life sciences, computing resources are an important part of the survival problem. Here we hope to get some guidance, new viewpoints, and suggestions from the sessions on networking and on hardware configurations.

Broadly, networking of computing certainly must be considered as having the potential to get us more computing of more different types, distributed to more investigators. If networking has not succeeded totally in the past—and indeed if our research computer dollar is ever shrinking—then perhaps we can spend one session trying to understand the computer networking problems and rewards.

The session of papers on computer hardware we can view similarly as offering at least the possibility to guide us to cost-savings by obtaining new classes of computer services for life science researchers. Your speakers are people who are deeply immersed in hardware development; it is not expected that most of us will emerge similarly expert. Most of us will want to increase our understanding of how the technical economies of the new configurations and the new circuit elements can ultimately be translated into cheaper and more satisfactory user services. If you become restive with the elucidations of timing improvements, then please press these speakers (and any others) to be certain that a new piece of hardware or arrangement of hardware does really improve user services or does really permit a new user insight into a life science problem. Some of the changes may merely be changes; some will be improvements.

It will not be our purpose to proselytize for minicomputers or hardware logic for special purpose dedicated hardware. The economic forces—and especially short term research funding programs (so it seems to me)—argue all too strongly for these smallish devices.

Counterpoised against these pressures is the problem created by our need for computer-searchable data bases—especially data bases with patient care data. If the medical record is ever truly to be the subject of research, if we are to be able to test our theories against the reality of patient treatment, then we must make, preserve, and search such data bases. These files are inevitably large, and inevitably (at least in the context of existing devices) must be searched by large computing systems. The group of papers on data base systems explores these problems. This is a world of minimum million-dollar-a-year budgets in which costs are allocated to a large number of users. You will hear of success and acceptance of these principles in industry and of tragic waste in the health care community. Some of what has been lost with losing big computers may be regained by skillful application of the new minicomputer techniques. Some formerly on-line data base inquiries may now be performed in an off-line mode. Some progress and initiative in the United States in patient data base building may simply be lost.

All of this is merely to say that we do what we can. Computing as a contribution to man's understanding of himself and his world is of major long-term importance. I personally believe that this will prove to be at least as powerful an advance as the understanding which came with the application of the microscope to the life sciences. Computing in biology and medicine simply must go forward. Indeed clearly it will go forward—in the United States, and at a quickening pace abroad.

We who use these machines may be able to increase our efficiency and reduce duplications and our other various faults through personal application of some or all of the principles and techniques we will hear of during this conference. Yet there is a level beyond that which we individually can achieve by simple belt tightening. It is this level that is to be addressed in the final session on meeting the costs. Costs, that is, of building systems and files to serve the nation, costs of training scientists to use and operate such schemes, costs to plan and administer growth in a comprehensible and orderly way. The session deals with a problem that is not merely a matter of beating the drums for more federal grants once again, nor merely of pleading for more consistency for long-range planning and research support—an end to on-again, off-again federal funding. All these things are certainly part of the solution to the cost problem. The greater part however must surely lie in a view of allocating the costs of proper computer support—at whatever level is appropriate—as a national cost. In any case it is this perspective we hope to hear in the last session.

In closing I would like to alert you to the occurrence of some informal sessions. The following foreign computer experts have agreed to meet with any participants to discuss their own systems and perspectives: Hans Peterson, M.D., Stockholm County, Medical Data Base System, Stockholm, Sweden; D. E. Clark, M.D., University of Manchester, Manchester, England; and Professor Christian Estrup, Instituttet for Kemiindustri, Denmark.

I hope each of you will have an enjoyable and intellectually prosperous time at the conference. If you have a problem, please tell me or Doctor Starkweather. Good luck and once again—welcome.

Session I

Computer applications I

Introduction

JULIA T. APTER, *Organizer and Chairperson*
Rush Medical College, Chicago, Illinois 60612

Computers are being used to help perform research by permitting computations not humanly possible otherwise, either because they would take too long or because they could not be done at all. Often, the results of such research make it possible finally to provide a service, either to other research projects, to teachers, or for patient services. Four of these presentations call attention to four such services and speak very well for themselves. The presentation by Dr. Robert Macey is of a different kind, placed in this session because it gives a basis for an interface between human thought or endeavor and electronic computers. It therefore introduces a clearer understanding of the entire conference.

That particular basis is currently spoken of as "modeling." In other words, biological experiments, ideas, or data are most properly formalized in a mathematical or physical form before computers can "understand" a biological problem or solve it. However, such formalization is required as much of physical problems as of biological problems and the kinds of formalization and their advantages currently being used can be appreciated from consideration of the following simple physical problem:

> "A cylindrical buoy, 12 inches in diameter and weighing 200 pounds, floats on the water with axis vertical. Find its period of oscillation, given that the upward force is equal to weight of the displaced water."

There are several ways I could find the answer.

By *Method 1,* I could go out to an actual buoy, hook up a number of sensors and transducers and record its movement which would appear as a series of small irregular oscillations induced by the wind, waves, or boats that pass by, as well as by the inadequacies of the attached sensors and transducers. But such measurements would fail to give the answer wanted and needed.

To get the proper answer, I would be required to use at least *Method 2:* to push the buoy down with a known force or displacement and *then* record the motion; (in other words, to distinguish between simple observation of uninfluenced motion and the behavior of the buoy during a suitable experiment). A computer could analyze the complex wave form produced by either *Method 1* or *2* as a Fourier transform, finding amplitude, frequency, damping factors, etc. In other words, the data of either *Method 1* or *2* would be fit to some curves whose parameters would make it possible to generate curves that resemble the data collected. The computer and this kind of model would help analyze data but would not give insight into the processes leading to the data. This kind of analysis would not predict, for example, how the buoy would behave in a storm. Using *Methods 1* or *2,* I would have to wait for a storm, measure the motion, and analyze it again, as before, with the help of digital or analog computers.

U.C.W. ABERYSTWYTH LIBRARY

By a *Method 3* it might be possible to quantify the correlation between the data obtained during a storm and the data obtained before a storm. This method would involve finding some transfer function that would transform the storm into its effect on the buoy. This is generally called "systems engineering" and it requires considerable data gathering without providing parameters characteristic of the system in question or generally predicting behavior to previously untested forcings.

For *Method 4,* a statistician would ask us to perform several experiments on differing buoys under differing conditions and then compute means, distributions, and deviations of the parameters taken from Fourier transforms and from transfer functions for each set of conditions. Although this process is strictly not modeling, but curve fitting, statisticians refer to it as "modeling" nevertheless.

A *Method 5* providing the answer to the buoy problem involves considering an idealized buoy, idealized water, and applying idealized physical laws, formalizing the problem into mathematical expression:

$$m\ddot{x} + b\dot{x} + kx = 0$$

where m would represent inertia, b energy dissipative, and k energy-conserving elements of the buoy–air–water system. This equation can easily be solved without a computer for the given situation if m, b, and k are known. By comparing it with the real motion of a real buoy, those parameters can be estimated ideally with an analog computer to solve all differential equations of the model and parameter estimation technique, and a digital computer to scan reasonable parameters until a set is found that makes the computer output match real motion. Such a model can also be used to predict behavior to an infinite variety of forcings like storms, passing boats, and even the initial displacement of the simplest experiment.

Many biological problems are being approached this way; Robert Macey tells us of some. Others can be found in my own works (1) which refer to still others. Many will be presented throughout the entire program. Be on the alert for them; it is very likely that they will be heuristic, helping those of you who currently approach biological problems mainly by *Methods 1* and/or *2*. Knowing how computers work may help you think of ways to model your particular research problems or to formalize your current "working hypotheses."

REFERENCE

Apter, J. T. Models and mathematics in medicine. In: *Medical Engineering,* edited by C. D. Ray. Chicago, Yearbook, 1974, 79–89.

Computer analysis of protein sequences[1]

M. O. DAYHOFF
National Biomedical Research Foundation
Georgetown University Medical Center
Washington, D.C. 20007

Most interesting of all of the biochemical components of living organisms are proteins and nucleic acids because of their potential for diversity in structure and function and their key positions in the evolutionary feedback system of each species. Sequence studies on these molecules have been carried out for various purposes in a number of disciplines: structural chemistry, crystallography, enzymology, genetics, evolution, systematics, physiology, immunology, pharmacology, genetic diseases, toxicology, virology, bacteriology, and even anthropology. In many of these disciplines, the protein and nucleic acid sequences and associated data are essential for a correct exposition of the fundamental concepts and for a parsimonious organization of the knowledge. A new field of study termed molecular evolution has emerged consisting of chemical, genetic, biological, and mathematical intercorrelations of such data. I will first describe the data available and then touch on some of the uses we make of the computer in research using this data.

The first protein structure, that of bovine insulin, was elucidated by Sanger in 1953. This was a unique structure, consisting of two chains made from 51 amino acid residues. It has since been shown that, with very minor exception, proteins are synthesized by complex cell machinery according to the message of a nucleic acid template. There are 4 kinds of nucleotides in the template and there are 20 kinds of amino acids, which are polymerized to make proteins through a chemical code. Some proteins are functional as initially polymerized; some undergo chemical modification: disulfide bonds are introduced, terminal groups or selected residues are modified, the chain is hydrolyzed, or small molecules may be bound to the protein. Occasional errors are made in the copying of the DNA from generation to generation. When such errors are beneficial or neutral, the new sequence may come to replace the original as the predominant form in the species.

We maintain a protein sequence reference data collection embodied in a series of books entitled *Atlas of Protein Sequence and Structure.* We have prepared a complete, currently correct volume of critically reviewed data and explanatory material each time the information has doubled. The first volume, with 44 sequences, appeared at the end of 1965 (7) and we published the fifth volume in September, 1972 (4). We anticipate another doubling in 5 years and have planned a series of supplements to appear in the intervening years. The first such supplement was published in June 1973 (5) and the second is in preparation. Our

[1] This work has been supported by NIH Grant GM-08710 from the National Institute of General Medical Sciences and by Contract NASW2546 from the National Aeronautics and Space Administration.

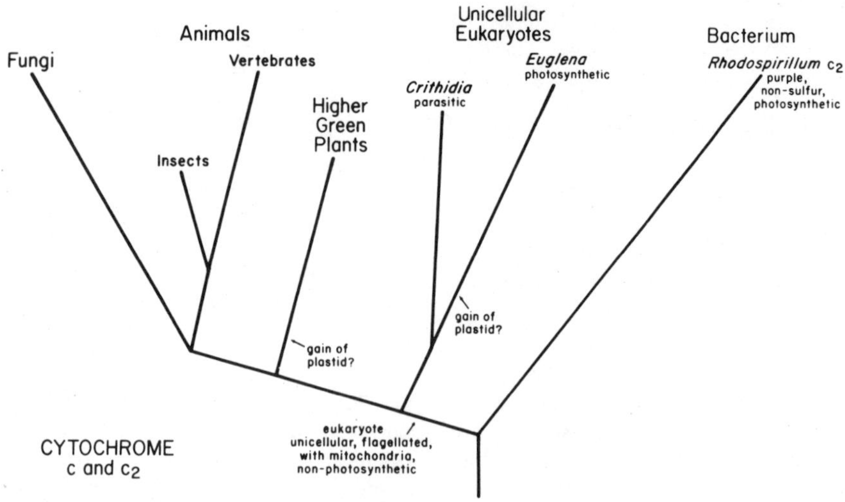

Figure 1. Outline evolutionary tree of cytochromes *c* and c_2. The evidence in cytochrome sequences indicates the order of divergence of eukaryote kingdoms shown here. The lengths of the branches are proportional to the number of mutations estimated to have occurred on these branches. The 'stem' of the tree was placed using earlier divergences of distantly related bacterial proteins. (Reproduced from Dayhoff, Barker and McLaughlin (6).)

scientific staff includes Dr. Lois T. Hunt, Dr. Winona C. Barker, Ms. Mary Jane Gantt and Mr. Menfai Shiu. We spend approximately one-third of our effort on research utilizing protein sequences, approximately one-third on data collection and research directly useful to the reference data center, and approximately one-third on preparation of manuscripts and computer-readable data bases and in performing services for others.

Over 2,500 experimental scientists have elucidated the 57,000 amino acid residues, derived from 619 sequences, currently on our sequence data tape. This is a small sample of the entire data base in living organisms, which may comprise some 10^{10} or 10^{11} different protein structures; in humans alone there are thought to be between 10^5 and 10^6.

As a first step in organization, we prepare alignments of families of sequences of similar structure and function. Proteins within a family usually differ at fewer than half of their amino acid positions and they are either homologues in various species or they are products of gene duplication accepted within the genome of one line; they have very similar names and functions. At present structures from 176 different families are known. For each family we routinely publish much auxiliary information on chemical reactivity, biological function, genetic details, and evolutionary history.

One of the most important uses of the computer is in determining from the sequences the evolutionary history of both the sequences and the organisms in which they occur. This is most conveniently represented by a tree. Determination of the best topology linking a set of sequences is a problem of double minimization of the inferred changes (tree size) with regard to the set of inferred nodal sequences and the topological configuration. The total of the differences between the various connected nodal or terminal sequences gives the tree size. First, a given topology is considered. An ancestral sequence is formed at each node and the changes inferred on each branch are counted, position by position, to give the total tree size. There is a single minimum size, which may correspond to more than one set of nodal sequences. Algorithms are available to determine these sets and the corresponding tree size (10, 14). The second step is to investigate all topologies and pick that topology with the minimum tree size. It is practical to consider all of the 105 structures of a tree of 6 sequences or even the 945 structures for a tree of 7. For larger trees we have developed programs to consider reasonable subsets.

The phylogeny of the kingdoms of higher organisms is derivable from cytochrome *c* sequence information as shown in Fig. 1 (6, 12). The development of physiological systems such as those involving thyroid and steroid hormone production will be illuminated by knowledge of the evolution of the proteins involved (see Fig. 2) (9). A combination of information from several proteins illuminates primate evolution and places human structures close to those of the chimpanzee and the gorilla (13).

Details of the changes inferred in the tree define the evolutionary process. From these changes we have developed a computer model of the point mutation

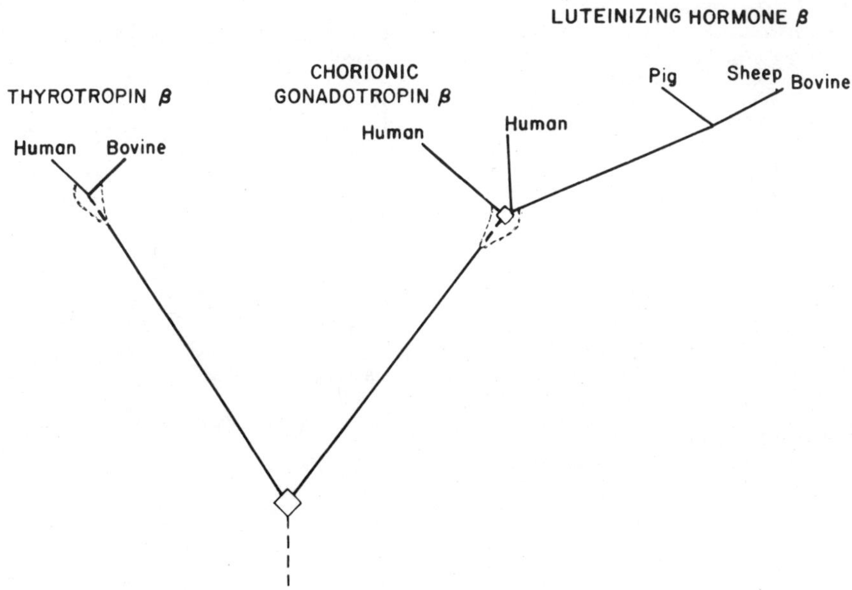

Figure 2. Evolutionary tree of the gonadotropin–thyrotropin beta chain superfamily. Thyrotropin (TSH) and luteinizing hormone (LH) are produced by the pituitary, while chorionic gonadotropin (CG) is produced by the fetal portion of the placenta. The branch lengths are drawn proportional to the amount of evolutionary change estimated to have occurred. Because the ancestral line between the TSH and the LH-CG tree is very long, one cannot place its attachment to the mammalian LH-CG tree with certainty. The connection fell within the length bracketed by dashed lines in ⅔ of the cases for simulated trees. Thus it is most likely that the gene duplication producing CG preceded the mammalian radiation. However, the gene seems to be expressed only in the primates. The point for the earliest gene duplication, represented by a diamond, was arbitrarily centered. We estimate that this earliest divergence occurred between 350 and 550 million years ago, early in vertebrate evolution. (Reproduced from Dayhoff, McLaughlin, Barker and Hunt (9).)

TABLE 1. Selected protein families and superfamilies

We have organized all of the presently known sequences into 176 families and 102 superfamilies. Proteins within a family usually differ at fewer than half of their amino acid positions and they are either homologues in various species or they are products of gene duplication; their similarity of function has usually been recognized before the sequence was known and they have identical or very similar names. These families can be further organized into superfamilies, where similarity of sequences in different families can be recognized by statistical methods and search procedures. Similarity, which is very unlikely to have arisen by chance, can be demonstrated where sequences are as much as 85–90% different. The families that comprise the more thoroughly investigated superfamilies are shown below.

Cytochrome c related
- Cytochrome *c*—Eukaryote
- Cytochrome c_2—*Rhodospirillum*
- Cytochrome c_{551}—*Pseudomonas*
- Cytochrome c_5—*Pseudomonas*
- Cytochrome c_6—Chloroplast
- Cytochrome c_{553}—*Desulfovibrio*
- Cytochrome c_{555}—*Chlorobium*

Ferredoxin related
- Ferredoxin—*Desulfovibrio*
- Ferredoxin—*Peptostreptococcus, Clostridium, Micrococcus*
- Ferredoxin—*Chromatium*
- Ferredoxin—*Chlorobium*
- Ferredoxin—Chloroplast
- Adrenodoxin—Mitochondrial
- Putidaredoxin—*Pseudomonas*

Trypsin related
- Trypsinogen—Animal
- Chymotrypsinogens A and B—Animal
- Elastase—Animal
- Thrombin B chain—Animal
- Alpha-lytic protease—Myxobacter

Basic protease inhibitor related
- Basic trypsin inhibitor—Mammalian
- Colostrum trypsin inhibitor—Mammalian
- Venom basic protease inhibitor—Viper
- Venom basic protease inhibitors—Mamba (Elapid)
- Isoinhibitor K—Snail

Globins
- Hemoglobin alpha chain—Vertebrate
- Hemoglobin beta-type chains—Vertebrate
- Myoglobin—Vertebrate
- Globin—Cyclostome
- Globin—Mollusc
- Globin—Insect
- Globin—Annelid
- Leghemoglobin—Plant

Thyrotropin beta chain related
- Thyrotropin beta chain—Mammalian
- Luteinizing hormone beta chain—Mammalian
- Chorionic gonadotropin beta chain—Mammalian

Glucagon related
- Glucagon—Vertebrate
- Gastric inhibitory polypeptide—Pig
- Secretin—Vertebrate
- Vasoactive intestinal peptide—Pig

Immunoglobulin related proteins
- Immunoglobulin lambda chain C region—Mammalian
- Immunoglobulin kappa chain C region—Mammalian
- Immunoglobulin gamma chain C region—Mammalian
- Immunoglobulin mu chain C region—Mammalian
- Immunoglobulin epsilon chain C region—Mammalian
- Beta $_2$-microglobulin—Mammalian
- Immunoglobulin lambda chain V region—Mammalian
- Immunoglobulin kappa chain V region—Mammalian
- Immunoglobulin heavy chain V region—Mammalian

process (8) characterized by a matrix of probabilities of mutation from each amino acid to each other one, incorporating also the relative probabilities that each given amino acid does not change at all. From this probability matrix we derive an odds matrix that can be used in comparing sequences for evolutionary similarity. Each term gives the odds that the pair of corresponding amino acids was derived through evolutionary replacement rather than by chance frequency of occurrence. This matrix has proven a powerful tool.

We routinely use several procedures based on the odds matrix. The first is a computer algorithm (2, 15) to score the alignment of two sequences and to compare this with the distribution of scores, after repeated randomizations of the two sequences. The probability that the real score is a member of the random distribution is obtained. A second method compares all fragments of a predetermined length from one sequence with all such fragments from the other (11). From this a distribution of thousands of scores is typically obtained of which at most only a few could be derived from fragments that are evolutionarily related. The average score of the upper part of this distribution is compared with the equivalent average scores from randomized sequences and again measures of significance are developed (1, 3). Finally, we search the known sequences for segments similar to a given segment. By these means we have organized the 176 families from which sequences are known into 102 superfamilies (see Table 1). Plant leghemoglobin is distantly related to the human hemoglobins and myoglobin, bacterial and mammalian trypsin show significant similarity, and mitochondrial adrenodoxin is related to plant chloroplast ferredoxin. Surprising results such as the high probability of relationship of nerve growth factor (5) to proinsulin or of epidermal growth factor to pancreatic secretory trypsin inhibitor (11a) are sometimes obtained.

TABLE 2. Average rates of mutation acceptance

Proteins	Accepted point mutations per 100 residues per 100 million years
Amyloid A	48
Growth hormone	35
Immunoglobulin C and V regions	32
Luteinizing hormone beta chain	30
Hemoglobin chains	14
Thyrotropin beta chain	8
Basic protease inhibitor	<6
Trypsinogen	5
Cytochrome *c*	3
Glucagon	1.1
Histone IV	0.06

Average rates for all families of proteins investigated range between those of amyloid A and histone IV. So slow is the change by point mutation that we could hope to detect proteins related to those in the bottom half of the table if they occur in bacteria or in higher organisms. The observed rates of amino acid mutation acceptance have been corrected for superimposed mutations. In calculating the average rates for most families, we used the difference accumulated since the divergence between the mammalian orders at 75 million years ago. For some proteins we used the estimated divergence times of other lines: amyloid A proteins from man and rhesus monkey, 20 mya; trypsinogen from dogfish and mammals, 400 mya; glucagon from birds and mammals, 300 mya; histone IV from plants and animals, 1,500 mya.

Proteins that change mainly by point mutations have been strongly conserved in evolution (see Table 2). So slow is this change that in living vertebrates homologues of all such proteins that were present in the ancestral vertebrate would still be recognized as members of the same protein superfamily. Primitive proteins within some superfamilies have changed so slowly that members would be recognized in the whole world of living creatures. Although the total number of different protein families found in living forms may number some 10^5, the number of superfamilies may be fewer than 1,000. The 102 superfamilies from which structures are presently known may thus represent 10% of the total.

There is much work yet to be done in understanding the data and in substituting logical processes for vast stores of undigested information. For chemistry, we need rules to predict the folding of the chain into a functional molecule, and we need to understand the evolution of metabolic capacity within the cell. For biology, we need to know the evolutionary history of biological species, the developmental history of individuals and the biochemical microanatomy of cells. For genetics, we need to know what the human genetic complement is and how it is organized and controlled. For mathematics, we need a formulation of information content, and of the efficiency of organization and function.

The key to a great advance in understanding in the life sciences is knowledge about all of the protein superfamilies through which living things are organized and controlled. It is our hope and expectation that this basic information will permit a simplification in exposition and organization, which will be significantly aided by the computer.

REFERENCES

1. Barker, W. C., and M. O. Dayhoff. Detection of homology between remotely related proteins. *Biophys. Soc. Abs.* 10: 152a, 1970.
2. Barker, W. C., and M. O. Dayhoff. Detecting distant relationships: computer methods and results. In: *Atlas of Protein Sequence and Structure 1972,* edited by M. O. Dayhoff. Washington, D.C.: Natl. Biomed. Res. Foundation, 1972, vol. 5, p. 101.
3. Barker, W. C., and M. O. Dayhoff. Computer studies of distantly related protein sequences. *Biophys. Soc. Abs.* 13: 205a, 1973.
4. Dayhoff, M. O., editor. *Atlas of Protein Sequence and Structure 1972.* Washington, D.C.: Natl. Biomed. Res. Foundation, 1972.
5. Dayhoff, M. O., editor. *Atlas of Protein Sequence and Structure, Volume 5, Supplement I 1973.* Washington, D.C.: Natl. Biomed. Res. Foundation, 1973.
6. Dayhoff, M. O., W. C. Barker and P. J. McLaughlin. Inferences from protein and nucleic acid sequences: early molecular evolution, divergence of kingdoms and rates of change. *Origins of Life* 5: 311, 1974.
7. Dayhoff, M. O., R. V. Eck, M. A. Chang and M. R. Sochard. *Atlas of Protein Sequence and Structure 1965.* Silver Spring, Md.: Natl. Biomed. Res. Foundation, 1965.
8. Dayhoff, M. O., R. V. Eck, and C. M. Park. A model of evolutionary change in proteins. In: *Atlas of Protein Sequence and Structure 1972,* edited by M. O. Dayhoff. Washington, D.C.: Natl. Biomed. Res. Foundation, 1972, vol. 5, p. 89.
9. Dayhoff, M. O., P. J. McLaughlin, W. C. Barker and L. T. Hunt. Evolution of sequences within protein superfamilies. *Naturwissenschaften* In press.
10. Dayhoff, M. O., C. M. Park and P. J. McLaughlin. Building a phylogenetic tree. In: *Atlas of Protein Sequence and Structure 1972,* edited by M. O. Dayhoff. Washington, D. C.: Natl. Biomed. Res. Foundation, 1972, vol. 5, p. 7.
11. Fitch, W. M. Further improvements in the method of testing for evolutionary homology among proteins. *J. Mol. Biol.* 49: 1, 1970.
11a. Hunt, L. T., W. C. Barker and M. O. Dayhoff. Epidermal growth factor: Internal duplication and probable relationship to pancreatic secretory trypsin inhibitor. *Biochem. Biophys. Res. Commun.* 60: 1020, 1974.
12. McLaughlin, P. J., and M. O. Dayhoff. Eukaryote evolution: a view based on cytochrome *c* sequence data. *J. Mol. Evol.* 2: 99, 1973.
13. McLaughlin, P. J., L. T. Hunt and M. O. Dayhoff. Techniques for determining protein sequence evolution applied to primates. *J. Human Evol.* 1: 565, 1972.
14. Moore, G. W., J. Barnabas and M. Goodman. A method for constructing maximum parsimony ancestral amino acid sequences on a given network. *J. Theoret. Biol.* Vol. 38: 459, 1973.
15. Needleman, S. B., and C. D. Wunsch. A general method applicable to the search for similarities in the amino acid sequence of two proteins. *J. Mol. Biol.* 48: 443, 1970.

Computerized clinical diagnosis

HOWARD L. BLEICH
Harvard Medical School at the Beth Israel Hospital
Boston, Massachusetts 02215

A time-shared computer has been programmed to assist the physician in the evaluation of body fluid and electrolyte disturbances (1, 2, 3). The program accepts clinical and laboratory data and generates in return an evaluation note designed to review the pathophysiology of the disorders with the physician and to assist him in their management. The purpose of the program is to provide the physician with on-the-job education in a form considered to be useful in solving selected clinical problems.

The program is written in the MIIS dialect of MUMPS, a string-manipulating interpretive language that runs on a PDP-15 computer. The program consists of two sections, one containing instructions to the computer and the other containing a bank of English text. The instructions collect input data, perform calculations, and select and arrange appropriate text strings that are then interleaved with excerpts from the patient's data and printed in the form of an evaluation note.

Distributed throughout the text are inserts that permit the program to modify the evaluation note for a particular patient. Insertion of given or calculated numbers permits adjustments of dosages. Insertion of words such as "relatively" and "probably" shades the meaning of borderline conclusions; other insertions may substantially alter or completely reverse the meaning of the text string. The purpose of the program is to act as a personalized textbook that presents to the physician only those excerpts from the medical literature that pertain to his patient, and to potentiate the usefulness of this information by interleaving it with given or derived information from the data base.

When the terminal is activated and the program is called, the computer transmits "electrolyte and acid–base evaluation," the date and the time; it then asks for the serum electrolytes in mEq/l (Fig. 1). After typing "Na =" the computer pauses while the user makes a response. If he knows the serum sodium concentration, he types the number and then pushes an enter button, whereupon the program advances to a request for the potassium concentration. If he does not know the potassium concentration, the user pushes only the enter button. If he makes a mistake in typing the value, he can erase with a rub-out button. If he gives a value that is clearly incompatible with life, the program will ignore it and repeat its request.

When sufficient information has been provided, the program computes the concentration of unmeasured anions, defined as the sum of the concentrations of sodium and potassium, minus those of chloride and total carbon dioxide content. If the anion gap is too small, a polite comment calls attention to the incompatibility and requests the electrolytes again. If the anion gap is acceptable, the program proceeds. If blood pH is provided,

```
ELECTROLYTE AND ACID-BASE EVALUATION   APR 18, 1974   1:39 PM

SERUM ELECTROLYTES (MEQ/L):

    NA     =  142
    K      =  2.8
    CL     =  92
    CO2T   =  36

BLOOD PH   =  7.48

BLOOD PCO2  =  48 (CALCULATED FROM CO2T AND PH)

PATIENT'S WEIGHT (IN POUNDS)     =  130

IS THERE CLINICAL OR RADIOLOGIC EVIDENCE OF CONGESTIVE HEART
FAILURE?...   NO

IS THERE PITTING EDEMA?...   NO

IS THERE EVIDENCE OF DEPLETION OF EXTRACELLULAR FLUID VOLUME?... NO

SERUM CREATININE (MG%)  =  1.2

BLOOD SUGAR (MG%)  =  112

24 HOUR URINARY POTASSIUM EXCRETION (MEQ/DAY)  =  56

HAS HYPOKALEMIA BEEN PRESENT FOR LONGER THAN 2 WEEKS?...   YES

DOES THE PATIENT HAVE SEVERE LIVER DISEASE?...   NO
```

Figure 1. Entry of patient's data. Underlined information was typed by the physician, everything else by the computer. Not shown in the figure are ways in which the program processes entries that violate required syntax, that are incompatible with life, or inconsistent with information entered previously.

pH, carbon dioxide content and a form of the Henderson–Hasselbalch equation are used to calculate the partial pressure of carbon dioxide. If the user enters a value that is not in close agreement with this calculated value, the program makes an explanatory comment and asks for resolution of the discrepancy. If, on the other hand, he presses only the enter button, which would indicate that he has not measured carbon dioxide pressure, the computer uses the space to print its calculated value.

The program then characterizes the type and severity of each component of the electrolyte and acid–base disorder. It then asks questions concerning renal function and extracellular fluid volume and whatever additional questions are needed to determine proper therapy. As soon as its demands have been met, the program prints its analysis (Fig. 2, 3). The evaluation begins by defining the components of the electrolyte and acid–base disorders. It then provides a list of diagnostic possibilities, a brief discussion of suggested therapeutic measures, and a small number of pertinent references to the medical literature.

In its present form, the program processes data concerning any combination of electrolyte and acid–base disorders. The data can be presented spontaneously in any order desired, or each piece of information can be entered when requested by the program. In either case, the data are checked for evidence of internal inconsistency, and additional information is requested as needed to solve the electrolyte and acid–base aspects of the clinical problem. If certain requested data are not available, the program is designed to draw the most useful conclusions possible from the data provided, to specify the limitations that pertain to these conclusions, to suggest diagnostic tests designed to circumvent these limitations, and, while awaiting the results, to suggest interim therapeutic measures. The time required to use the program at 110 baud is approximately 7 minutes.

The chemical model employed in this program is considerably more simple than the complex model of DeLand and Bradham (4), but is felt to be adequate for clinical and educational purposes. Our model also includes a great deal of clinical information which is not included in the DeLand model.

Experience in our hospital reveals a number of cases in which patient care has been favorably influenced by this program (2). On one occasion, physicians had decided to administer hypertonic saline to a hyponatremic patient when advised by the program to restrict fluids instead. Consultation was requested, and the latter course adopted. The kidney service has received numerous consultations after the program revealed a problem not previously appreciated by the physicians. At times, the program outperforms its inventor; it was, after all, written during lucid intervals, while rele-

```
                         EVALUATION NOTE

     THE PCO2 OF 48 MM HG AND PH OF 7.48 UNITS ARE COMPATIBLE WITH
UNCOMPLICATED METABOLIC ALKALOSIS. THE MOST COMMON CAUSES OF
METABOLIC ALKALOSIS ARE:
     1) VOMITING OR GASTRIC ASPIRATION
     2) DIURETIC THERAPY (THIAZIDES. MERCURIALS. ETHACRYNIC ACID. OR
        FUROSEMIDE)
     3) HYPERADRENALISM (CUSHINGS SYNDROME. PRIMARY OR SECONDARY
        ALDOSTERONISM. ADRENAL STEROID THERAPY) OR LICORICE
        INGESTION
     4) EXCESSIVE ALKALI INTAKE (ALKALOSIS USUALLY TRANSIENT)
     5) POST HYPERCAPNIA (PARTICULARLY IF CHLORIDE INTAKE IS SMALL)
     6) HYPERCALCEMIA (POSSIBLE--LITERATURE NOT CLEAR)
     7) BARTTER'S SYNDROME (A RARE DISORDER CHARACTERIZED BY
        HYPOKALEMIA. HYPOCHLOREMIA. METABOLIC ALKALOSIS. AND. IN
        SOME INSTANCES. HYPONATREMIA. SHORT STATURE. AND MENTAL
        RETARDATION)

     IF HYPERADRENALISM IS ABSENT IT SHOULD BE POSSIBLE TO CORRECT
THE METABOLIC ALKALOSIS BY ADMINISTERING SUFFICIENT CHLORIDE TO
REPLACE PREVIOUS LOSSES AND TO ALLOW A URINARY CHLORIDE EXCRETION
OF AT LEAST 10 TO 20 MEQ PER DAY.  IN AN EFFORT TO ACHIEVE THIS
GOAL IT IS SUGGESTED THAT. IN ADDITION TO REPLACING KNOWN CHLORIDE
LOSSES. 120 MEQ OF SODIUM. POTASSIUM. OR AMMONIUM CHLORIDE BE GIVEN
DURING THE NEXT 24 HOURS. AND THAT SERUM ELECTROLYTES AND BLOOD PH
BE MEASURED AT LEAST DAILY UNTIL SIGNIFICANT IMPROVEMENT IN THE
ACID-BASE ABNORMALITY OCCURS.
```

Figure 2. First section of evaluation note generated by the program immediately after the last entry was made in Fig. 1. Since the patient is an adult (weight = 130 pounds), the suggested dose of sodium bicarbonate (usually less than the estimated deficit) is rounded off to the nearest ampule. Note that the program does not attempt to make a diagnosis, but provides a differential list for consideration by the physician.

```
     THE FINDING OF A LOW SERUM POTASSIUM CONCENTRATION (K = 2.8
MEQ/L) INDICATES THAT BODY POTASSIUM STORES HAVE B EN SIGNIFICANTLY
DEPLETED.  FURTHERMORE, THE FINDING OF SIGNIFICANT URINARY
POTASSIUM EXCRETION (56 MEQ/DAY) IN A PATIENT WHOSE POTASSIUM
DEPLETION HAS BEEN PRESENT FOR AT LEAST 2 WEEKS INDICATES THAT
RENAL POTASSIUM CONSERVATION IS IMPAIRED.  COMMON CAUSES OF RENAL
POTASSIUM WASTING INCLUDE, IN ADDITION TO THE CAUSES OF METABOLIC
ALKALOSIS, OUTDATED TETRACYCLINE, RENAL TUBULAR ACIDOSIS (NOT
PRESENT HERE), AND THE DIURETIC PHASE OF ACUTE TUBULAR NECROSIS.

     IN AN EFFORT TO REPAIR THE POTASSIUM DEFICIT SIMULTANEOUSLY
WITH THE METABOLIC ALKALOSIS IT IS SUGGESTED THAT, IN ADDITION TO
REPLACING ANTICIPATED POTASSIUM LOSSES AT LEAST 6Ø MEQ OF THE
CHLORIDE RECOMMENDED ABOVE BE ADMINISTERED AS POTASSIUM CHLORIDE.
BECAUSE OF THE SEVERITY OF THE HYPOKALEMIA (K = 2.8 MEQ/L), THIS
SHOULD PROBABLY BE DONE WITHIN THE NEXT 12 HOURS AND SERUM
ELECTROLYTES MEASURED AGAIN AT THAT TIME.

     THANK YOU FOR REFERRING THIS INTERESTING PROBLEM TO US.

REFERENCES:

     1) SHEAR, L. AND BRANDMAN, I.S.
        HYPOXIA AND HYPERCAPNIA CAUSED BY RESPIRATORY COMPENSATION
        FOR METABOLIC ALKALOSIS
        AMER. REV. OF RESP. DIS.. 1Ø7,836,1973.
     2) BEESON, P.B. AND MCDERMOTT, W.
        CECIL-LOEB TEXTBOOK OF MEDICINE.
        13TH EDITION (1971), P.1628.
     3) GARELLA, S.. CHAZAN, J.A.. AND COHEN, J.J.
        SALINE-RESISTANT METABOLIC ALKALOSIS OR "CHLORIDE-WASTING
        NEPHROPATHY."
        ANN. INT. MED.. 73:31,197Ø.
     4) SCHWARTZ, W.B.
        POTASSIUM AND THE KIDNEY
        NEW ENG. J. MED.. 253,6Ø1,1955.
```

Figure 3. Remainder of evaluation note. References are listed in the order in which the medical decisions to which they pertain are found in the evaluation note.

vant literature was at hand and more knowledgeable physicians nearby.

The generation and continual revision of such a program demands that the actual process of clinical decision making be formulated with utmost precision. Such a precise formulation requires that the investigator clearly distinguish what is known from what is assumed; it thus serves as a stimulus for study of additional areas where clinical or laboratory research should be done.

Finally, this program was developed not to circumvent the thoughtful clinician, but rather to provide the student, house officer and practicing physician with a source of continual and stimulating instruction. The purpose has been to develop an educational tool, one which is as reliable as it is indefatigable, and which is particularly effective because it relates what it teaches to a current clinical problem.

REFERENCES

1. Bleich, H. L. Computer evaluation of acid-base disorders. *J. Clin. Invest.* 48: 1689, 1969.
2. Bleich, H. L. The computer as a consultant. *New Engl. J. Med.* 284: 141, 1971.
3. Bleich, H. L. Computer-based consultation. *Am. J. Med.* 53: 285, 1972.
4. DeLand, E. C., and G. B. Bradham. Fluid balance and electrolyte distribution in the human body. *Ann. N.Y. Acad. Sci.* 128: 795–809, 1966.

Biostatistical computing

CAROL M. NEWTON
University of California, Los Angeles, California 90024

The purposes of data analysis are the purposes of research itself—to make new discoveries and to establish their certainty to the extent that they may be entered into the body of firm scientific knowledge that is the foundation for future research and for the application of what we have learned. From this, two conclusions follow: First, data analysis should involve the researcher himself, albeit with the help of biostatisticians and other methodological experts. Second, data analysis for the sake of discovery should be valued as highly as data analysis for the validation of hypotheses. Already computers have brought to biomedical research unprecedented support for full utilization of hard-won experimental data, and they doubtless have played an important role in the increasing standards of quality for scientific reporting. But it is possible that the most exciting developments lie ahead, where new interactive techniques will increasingly involve the biomedical researcher himself and will contribute to evolution of the scientific method by providing innovative supports for the process of hypothesis discovery.

BACKGROUND

When man started to learn about the world he proliferated many hypotheses, much as we do now in our initial steps toward understanding an unknown phenomenon. The problem was that these competed in the realm of opinion rather than of scientific experimentation. Some exceptions were the early laws of physics. They rose with authority from data bases that had little inherent variability. But the growth of knowledge in more complex, inherently variable fields such as biology owes much to the dictates of the scientific method and to the statistical procedures and analyses it inspired. These reacted against the realm of opinion and emphasized reliable methods for the experimental validation of hypotheses. The power of analytical methods grew steadily. However, the effort and expense of implementing some inhibited their utilization for research. Not many years ago manual performance of a factor analysis that now might require far less than a minute on the computer was one of the major components of some dissertation efforts.

Needless to say, computers have had a variety of impacts on statistical supports to biomedical research. Most obviously, they have offered sophisticated statistical tools promptly and at little cost to all investigators. Relieved of time-consuming hand calculations, the consulting biostatistician has been able to serve far more investigators better and at less expense. He has had more opportunities to gain experience in the use of advanced statistical methods, and often more time to contribute to this advancement. New experiments can be designed with knowledge of results of preceding experiments, shortly after the latter have been completed. Thus, statistical supports have become infused far more widely into biomedical research, and this very likely has contributed to higher standards for acceptable scientific reporting.

WHAT IS AVAILABLE AT PRESENT

Dixon and Jennrich's recent review of statistical programs (2) provides basic information about 38 different systems. This information includes the type(s) of machine on which they run, the amount of core memory required, the language(s) in which they are programmed, the number of programs and subroutines in each set, some indication of the types of programs and documentation available, and people to whom one may write for further information. For 20 frequently used systems, there is a table indicating the presence or absence of important features such as whether observations may be weighted, data sets with missing elements can be accommodated, the user can specify the formats of his own data cards rather than having to conform them to the program's requirement, and so on.

It is important to note that these packages are organized in several different ways. Some, such as SSPs and the UNIVAC 1108 STAT PACK, are offered as sets of subroutines, the building blocks from which programs are made. These facilitate the development of programs especially suited to a given user. Further programming is not necessary for packages of complete programs, such as the BMD series of statistical programs which was developed under NIH support especially for biomedical users. At most, the user must supply brief FORTRAN descriptions of functions he wants to fit to data or of where data are located on his cards or tapes, and he is assisted in doing so by instructions in the BMD Manual. Statistical compilers such as ASCOP or DATA-TEXT, as well as multiple-purpose programs such as P-STAT, OMNITAB, and SPSS, facilitate the user's ability to direct a number of quite different analyses to his data during submission of a single job to the computer. Obviously, each approach has its advantages. However, the beginning user may prefer the second; he is not required to program and he can concentrate on how to accomplish the single analysis he first wants, moving on to sequences of different analyses later. He can accomplish these sequences with single programs by generating tape records of intermediate results. As his confidence and capabilities develop, he then may wish to try using the statistical compilers or mulitple-purpose programs, or to develop some programs of his own with the help of subroutine packages or subroutines available in the system he already is using.

Most computing centers associated with universities or other large institutions have statistical programs in their libraries. BMD programs are among the most widely distributed (1). They are known to have been adapted to machines other than those listed in the aforementioned reference. Information concerning this, as well as tapes containing the programs, can be acquired from the BMD Program Librarian, Health Sciences Computing Facility, School of Medicine, UCLA, Los Angeles, California 90024. BMD's 71 programs cover the categories of data description and screening (e.g., cross tabulations, histograms, plots), regression (fitting of curves to data), analysis of variance, multivariate analysis (e.g., factor analysis, discriminant analysis, clustering), time series analysis, nonparametrics, and utilities (e. g., sorting, data transformations). If some of these names are unfamiliar to you, consultation with a biostatistician might reveal that you could be extracting far more information from your data or designing more effective experiments with statistical aids such as these at hand.

ELEMENTS IN ACCESS TO STATISTICAL COMPUTING

Awareness

An ultimate barrier to use of any aid is lack of awareness of what it can contribute and how it can be procured. An increasing awareness of the roles of statistical supports results from refereeing, project reviews, scientific reading, the wider availability of biostatistical consultants, and the observation that some people are extracting more information from data equivalent to ones own. Yet, while there appears to be a growing tendency to include some biostatistical training in medical curricula, not many biological or medical sciences require such of their doctoral students. One very likely reason is that few appropriate courses have been developed for them, courses that include an interrelated blend of computer, statistical, and modeling methods with primary emphasis on their application to real biomedical problems. To be acceptable in crowded graduate curricula, such courses should be highly efficient, minimize mathematical derivations, use computers for most homework, and reach beyond what the student can do himself, to make him aware of the additional powerful supports a biostatistical colleague can provide him in both the design and analysis of experiments.

Data capture

Researchers will tend to agree that the most arduous aspect of using computers, as seen by them, is getting their data in a form that can be used by computers. This also is likely to be the most costly part of the process. Because most of the computers used today are digital computers, computers that handle discrete numbers or sets of characters, continuous physiological signals such as electrocardiograms must first be processed through an analog-to-digital converter, a device that delivers a sequence of numbers representing voltage as a function of time. This also may require intermediate recording, filtering or other signal conditioning, and the handling of tapes. Even when data are digital, they often must be transcribed to data-acquistion forms for keypunching, and occasionally it is necessary to look up numerical codes for diagnoses and such. However, once the data are in the computer, it is easy and relatively inexpensive to have automatic checks on the data's completeness, consistency, and ability to meet specified bounds checks. As the use of time-shared terminals becomes more widespread, the data input process should be substantially facilitated by conversational programs that check data immediately, providing a verifiable automatic coding for diagnoses and action-oriented messages when errors have been detected. In any case, consultation with an expert in biomedical computing is strongly recommended to design a data capture system that will be economical and easy to use.

Computer availability

Small, useful statistical calculations can be done on a number of the mini-computers that now are common in research environments. However, the user often must write his own programs or employ those that derive casually from users' groups or friends. Because the time-tested, widely available statistical programs described earlier are offered primarily on larger digital computers, access to them will be stressed in this section.

Ten years ago, many biomedical research communities did not have access to large computers. They had to make do with small computers shared with hospital accounting services, beg costly access to machines dominated by the physical sciences and engineering, or resort to remote job entry over telephone lines to distant biomedically oriented computer centers. At that time both telecommunication systems and the computers they accessed were not steadfastly reliable. Furthermore, in their novelty, computers then were regarded as status symbols rather than as the dumb beasts of burden they now happily are recognized to be. So then some people felt it demeaning to use another person's computer, whereas now they sensibly accept it as their due for fees paid. Modern time-sharing systems, where a number of local and remote users simultaneously and economically access a large computer, are largely responsible for important changes in what may be offered the individual researcher, and hence in his attitudes toward sharing a resource.

However, vestiges of old problems remain and in some ways are intensified. Committees setting policy for university computing centers still tend to draw heavily on mathematicians, the physical scientists, and engineers. And now the financial problems of some university computing centers are so dire that faculty are coerced to use them rather than alternatives that are clearly preferable on an individual basis. The funded biomedical investigator is an especially visible target for such coercion. Although remote biomedical computing centers that would provide him especially appropriate computer services, programs, and expert consulting to support his local biostatisticians very likely would be the most cost-effective solution for many of his more sophisticated needs, administrative impediments may prove to be more stubborn than technical difficulties in realizing such a solution.

Statistical programs (software)

No statistical package has all of the features and types of programs one might conceivably wish. Selectivity in package design must be exercised in the interests of economy, ease of use, and so on, and tends to reflect priorities most frequently expressed in the population of users for which the system is intended. The BMD programs have evolved in an active biomedical consulting and research program.

Consulting

Perhaps the greatest need of all is for appropriately trained biostatistical and biomathematical consultants, for people who effectively combine high-level skills in mathematics, statistics, and computing with substantial training or considerable working experience in biology and medicine. These people are rare. The need for them is extensive and great. One such person can markedly upgrade or facilitate the research of many biomedical colleagues. Appropriate training is demanding upon the student and upon those providing it. Before they were peremptorily cut off, Public Health Service training programs were supplying some such professionals, but far, far fewer than are required. As the public interest demands restoration of training programs, there is a danger that the focus will be on more familiar and dramatic categories such as heart and cancer, rather than on the training of skilled professionals whose mission is the general advancement of biomedical research through consulting, basic research on more powerful biomathematical and biostatistical supports, and the development of computer programs and systems that aid the effective infusion of new techniques into biological research and health care.

The concept of resource sharing should extend beyond computers, to consultants. Local "primary-care" consultants should be able to access expert specialists at the larger biostatistical computing centers and elsewhere. Our facility is developing tandem-terminal software for this purpose.

THE FUTURE

There is time to mention only two of a number of very interesting trends.

Computer-aided statistical research

Often computer approaches can prevail where exact mathematical developments fail. The importance of biomedical research, together with the difficult problems and data bases encountered there, has forced the development of statistical techniques that reach beyond what can be based on exact mathematical derivations. Instead, carefully designed simulation experiments upon the computer may be used to validate the new statistical approaches to stated levels of significance for

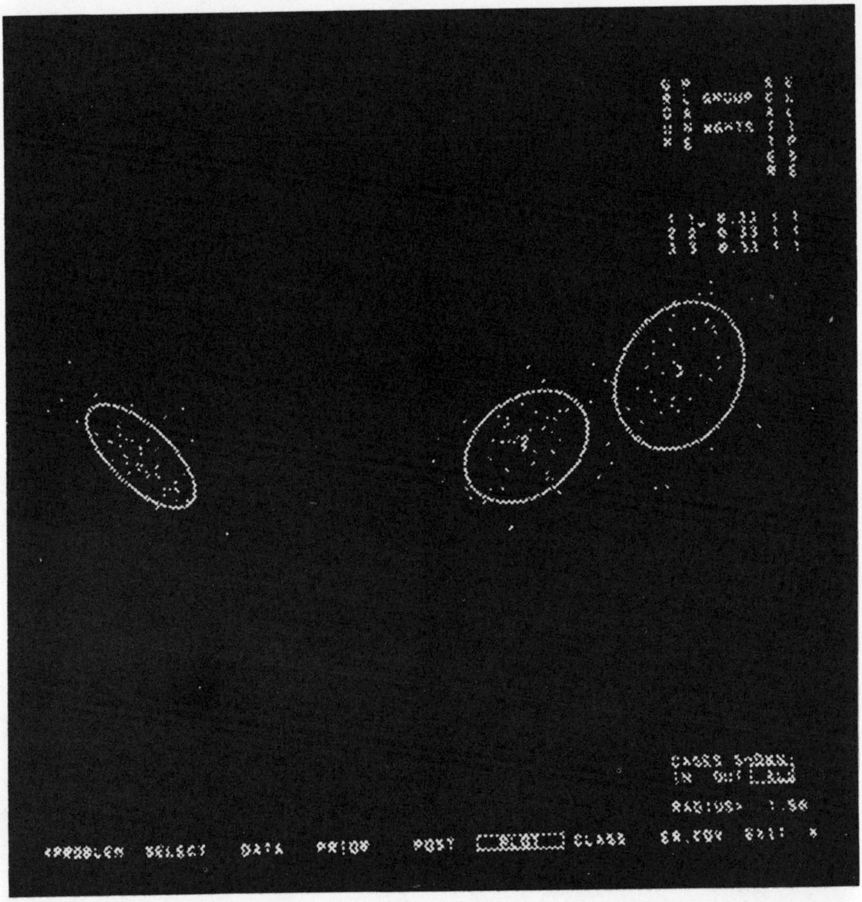

Figure 1. Several groups of multivariate observations appear as clouds of points. The dots are the projections of individual points on the plane determined by the centroids of any three groups. The ellipses are centered on the centroids of each group and have direction and size corresponding to the scatter of points about the centroid in the stated projection plane.

specified types of applications. Some people include this in the New Statistics. Interest in it is developing rapidly and is encountered as well in areas other than biostatistics. Established methods are being questioned; e.g., in some cases, least-1.7-power fits to data seem to have better properties than traditional least-squares fits. The prospects for this new area of research and for its beneficiaries are exciting.

Interactive graphics

At the outset it was noted that since data analysis is at the heart of all scientific research, the researcher himself should be involved in it, and it should relate as much to the discovery of new hypotheses as to their validation. In its early support to scientific methodology, statistics has emphasized validation. Computers have enabled the more computationally demanding of these supports to be routinely used throughout scientific research. Now one wishes to direct statistical computing toward two new highly related goals, supports to scientific discovery and devices for involving the researcher himself in data examination.

Already programs, such as the BMDs, that facilitate generation of a variety of graphs, tables and simple descriptive statistics, have encouraged researchers to look for unexpected relationships within their data, and more advanced multivariate programs have directed their statistical colleagues to others. One wishes to freely employ a combination of these while searching for relationships within data. Also, when one is searching for new insights rather than performing a rigorous validation, flexible manipulation of the data base should be enabled. Does removal of some possible outliers wipe out a suspected relationship? What happens when only cases having a certain combination of properties are considered? What do different clustering or discriminant approaches reveal when applied to the data (Fig. 1)? What functions or other models appear to fit best (Fig. 2)? Do new

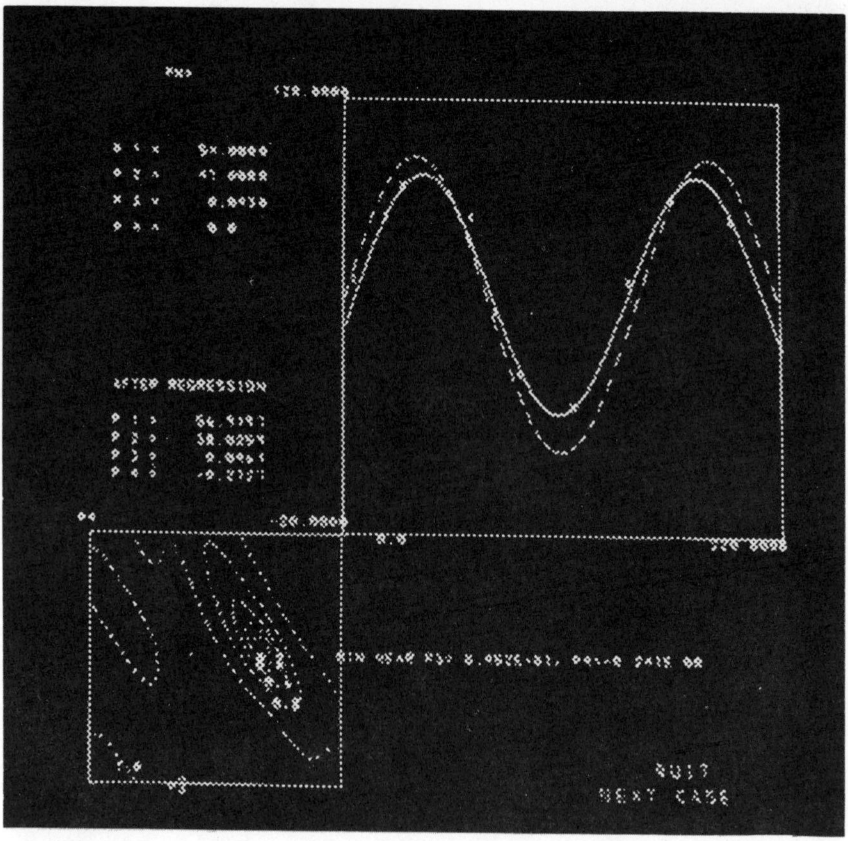

Figure 2. In this graphics adaptation of BMDP3R, different nonlinear models can be fitted to data. The dashed curve in the upper-right display relates to initial parameter estimates, and the solid curve, to the fitted parameters. The lower-left contour plot depicts the error term, the summed squares of residuals, for user-specified ranges of any two of the fitted parameters.

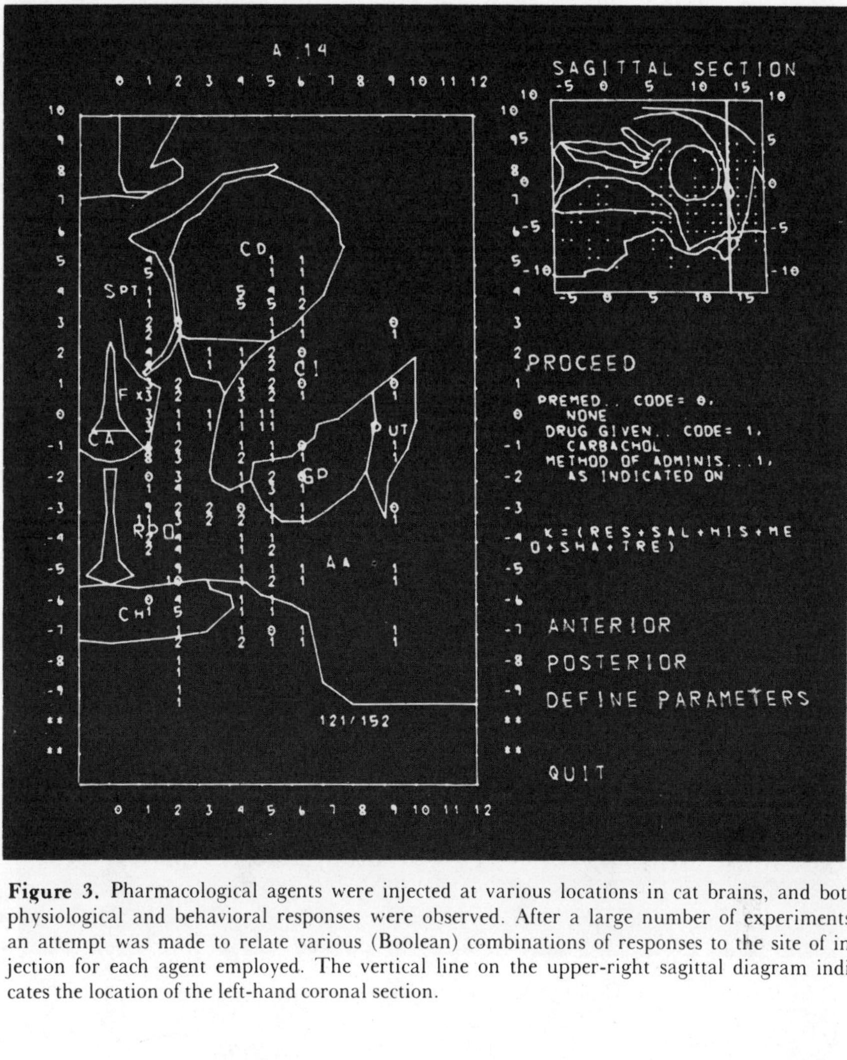

Figure 3. Pharmacological agents were injected at various locations in cat brains, and both physiological and behavioral responses were observed. After a large number of experiments, an attempt was made to relate various (Boolean) combinations of responses to the site of injection for each agent employed. The vertical line on the upper-right sagittal diagram indicates the location of the left-hand coronal section.

relationships become apparent when data are related to appropriate anatomical (Fig. 3) or geographical (Fig. 4) maps? One could continue at length, but the point surely has been made that flexible, versatile, graphically supported, interactive approaches are required.

The same interactive graphics devices that brought us strikingly three-dimensional representations of rotating molecules (3) can be effectively applied to visualization of three-dimensional projections of data. But now equivalent interactive graphics terminals have fallen in price from $100,000 to $250,000 down to $17,000, and some useful terminals can be purchased for less than $10,000. Furthermore, some of these terminals can access host computers over ordinary dial-up telephone lines and can be programmed in extensions of FORTRAN (5). The latter feature facilitates construction of interactive graphics data analysis software by use of existing FORTRAN

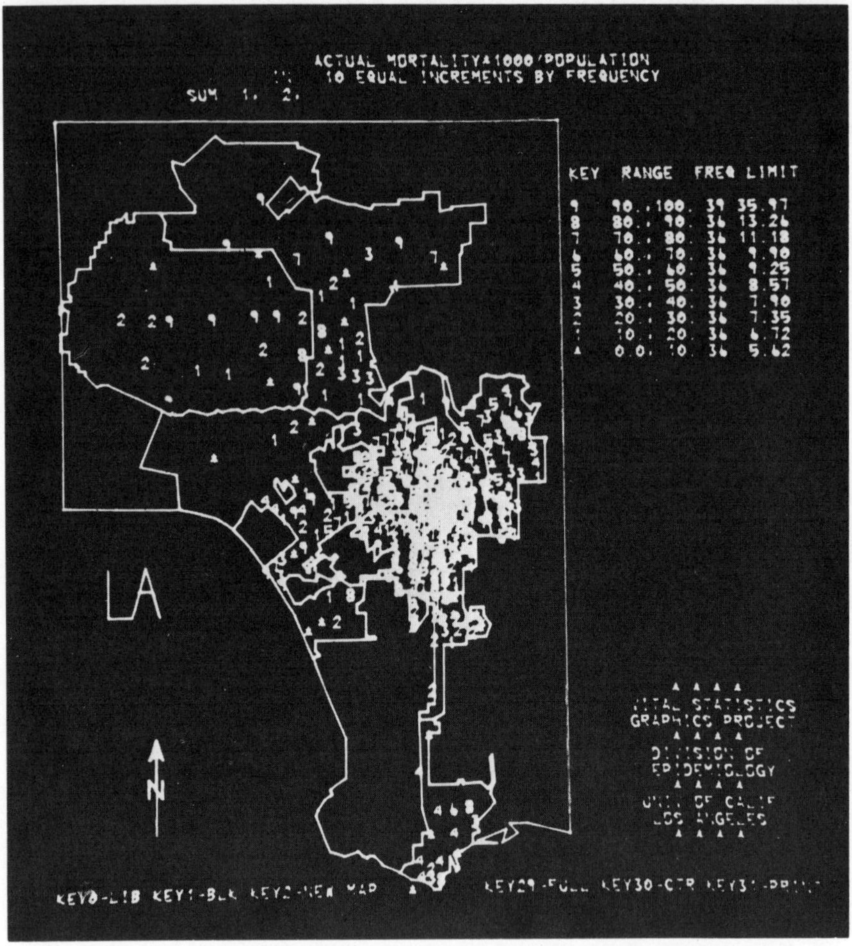

Figure 4. Census and heart disease data are related to a map of Los Angeles County.

statistical subroutines, which are widely available.

We now should ask ourselves what ideally is required to facilitate scientific discovery, to support the researcher's intuition and his effective collaboration with biomathematical and biostatistical colleagues. We should reexamine some ideas that may have been set aside earlier because of difficulty of implementation. For instance, would Sterling and Pollack's robot data-processing techniques (6) now be more appealing if implemented as interactive aids to the investigator? For all of the drama and beauty of scientific models, it must be conceded that creative data analysis often has been more fundamental to major scientific breakthroughs. Recall the discovery and consequences of the Periodic Table. But ideally modeling and data exploration should be pursued hand-in-hand. With modern interactive graphics techniques, this is quite possible (4).

REFERENCES

1. DIXON, W. J., editor. *BMD Biomedical Computer Programs*. Berkeley: Univ. of California Press, 1973.
2. DIXON, W. J., AND R. L. JENNRICH. Scope, impact, and status of packaged statistical programs. *Ann. Rev. Biophys. Bioeng.* 1: 505, 1972.
3. LEVINTHAL, C. Molecular model building by computer. *Sci. Am.* 214: 42, 1969.
4. NEWTON, C. M. Graphics in medicine and biology. *Proc. Natl. Computer Conf.*, 639–642, 1973.
5. RYDEN, K. H., AND C. M. NEWTON. Graphics software for remote terminals and their use in radiation treatment planning. *Proc. Spring Joint Computer Conf.* 40: 1145, 1972.
6. STERLING, T. D., AND S. V. POLLACK. *Introduction to Statistical Data Processing*. Englewood Cliffs, N. J.: Prentice-Hall, 1968, Chapters 10 and 12.

Mathematical models and membrane permeability[1]

ROBERT I. MACEY AND LENORE T. WADZINSKI
University of California, Berkeley, California 94720

At one time applications of mathematical models to physiology and medicine were largely recorded in a few esoteric journals read primarily by specialists who were involved in their development. Most laboratory biologists were neither interested nor had the necessary background to exploit them. Today, through improved educational curricula and through the introduction of modern computers, the situation is radically different. Mathematical models reach into virtually every field of biological inquiry. In many instances, they are no longer a luxury afforded to a few theoreticians; they have become an indispensable tool for the practical design, execution, and interpretation of experiments.

Asking for a model is simply another way of asking "How does it work?" Computers begin to play a role when models become formalized usually by the introduction of mathematics. Mathematical models can offer several advantages to the laboratory investigator:

1) Mathematical models can be used to coordinate a large amount of complex data. In this sense they may be a simple shorthand for what is already known. If the model is particularly well developed, then it may serve in lieu of experiments. Instead of making costly measurements, you need only calculate. In the life sciences, this stage of development is often more of an aspiration than a realization, but there are instances, e.g., the Hodgkin-Huxley model of the axon (2, 5), where nontrivial applications of this type are at hand.

2) Even if the model cannot substitute for experiments, models are very often enormously useful in the design of laboratory experiments. They provide the investigator with estimates of where to look, how long to wait, what to expect.

3) Models can also be used to simulate experiments that cannot be done in the laboratory.

4) Models often provide the only method for extracting characteristic parameters, e.g., binding constants, secretion rates, permeability coefficients, from experimental data.

5) Finally, models provide a potential means for discovery of qualitatively new phenomena. The fact that the nonpropagated membrane action potential in squid axons does not behave in an all-or-none fashion is a good example of discovery by computer simulation (3). In particular, the computer simulation revealed that the phenomena would only be apparent within a certain temperature range and this information led to the experimental demonstration of graded responses.

In this article we illustrate some of these advantages with examples taken from experimental membrane physiology. The general problem of membrane physiology is to relate transport properties to

[1] This work was supported by National Institutes of Health Grant GM-18819-03.

membrane structure. Most popular versions of membrane structure assume that the plasma membrane is composed of a lipid bilayer background with protein sometimes confined to the surfaces and sometimes penetrating through the bilayer. This suggests at least two transport pathways: *1*) through the lipid bilayer and *2*) through the protein portions. Passage through the lipid portion implies that permeation of lipid-soluble solutes will be rapid and this has been confirmed many times; description and interpretation of permeability coefficients for lipid-soluble substances is in a fairly satisfactory state (4). Transport of water and small polar solutes is not easily explained in these terms and as a consequence the existence of water channels or pores probably formed by polar portions of the protein is often postulated.

Pores provide a simple explanation of the dependence of permeability on molecular size. However, the chief evidence for their existence comes from studies of water permeability. The most impressive result of these studies is the demonstration that measurements of water permeability under an osmotic or hydrostatic pressure gradient cannot be utilized to predict the diffusional permeability measured with tracer water (9). This discrepancy is reconciled by the pore hypothesis. If for simplicity we assume continuum mechanics is applicable to the pore, then hydraulic or osmotic flow through the membrane should be proportional to the fourth power of the pore radius while diffusion through the pore should be proportional to its area, or the square of the radius. Hence diffusional and hydraulic permeabilities cannot be brought into agreement unless the pore radius is known. If this interpretation is applied to human red cells, it predicts a pore diameter in the neighborhood of 8 Å—a value that is similar in magnitude to the diameter of a water molecule. Thus our literal interpretation leads to an inconsistency; the assumed continuum mechanics predicts a pore radius that is too small for continuum mechanics to apply. Theoretical problems of hydraulic flow in narrow channels have not been solved, but recently they have been attacked by David Levitt in a computer simulation (7). Levitt's method was to assume a pore of a given size, place a given number of molecules (hard spheres) inside the pore, and assign a random position and velocity to each molecule. The computer then calculated the subsequent history of each molecule, i.e., its path as it moves about interacting with other molecules and with the walls. His results showed surprising agreement with continuum mechanics; the simulation predictions for the pore radius, for example, corresponded rather closely with the pore radius calculated on the basis of continuum mechanics. This study shows how a significant computer simulation need not involve a "mathematical" model. Further it shows the use of a computer to simulate an experiment that could not be done in the laboratory, and finally it is an example of unanticipated findings—a new "experimental" result.

On the whole, the evidence for water channels seems to be fairly good; it is the best simple explanation for available data. If we accept existence of these channels, the question of which solutes use them as a primary means of transport remains to be answered.

Our own interest in this problem arose when we found that we were able to open and close these pores in human red blood cells by judicious use of sulfhydryl reagents (8). When we studied the transport of small solutes we found a class of polar solutes, urea and related compounds, whose permeability was decreased with the pores closed. This was the expected result. However, we found a much larger class which showed no change and some even increased their permeability when the pores were closed. It appeared as though we had established a criterion for deciding which solutes were transported via pores. But further studies with the reagent phloretin showed that it could reduce the permeability of urea by a factor of 50 without affecting water permeability in the slightest. Since both phloretin and sulfhydryl reagents are known to inhibit carrier systems, the result suggested that urea was transported primarily by a carrier. This has important consequences for the pore theory. It suggests that water channels transport water and nothing else.

Resolution of the problem is possible only by detailed study of permeabilities of penetrants like urea which show changes when the pores are closed. Unfortunately, these solutes penetrate cell membranes very rapidly and a direct measurement of their permeation is difficult. A common indirect method for measuring the permeability of urea is to expose cells to a sudden increase in urea concentration and follow changes in cell volume with time. The problem is to relate the cell volume changes to the permeabilities of urea and water so that these parameters can be measured. This is accomplished by the mathematical model developed by Kedem and Katchalsky (6). Letting L_p and P_s represent osmotic and solute permeability coefficients, and letting σ denote the reflection coefficient of permeable solute, then the volume flux J_v and solute flux J_s per unit area of membrane can be written in terms of relevant concentration gradients as:

$$J_v = -L_p(RT\Delta C_m + \sigma RT\Delta C_s) \qquad (1)$$

$$J_s = (1 - \sigma)\bar{C}_s J_v + P_s \Delta C_s \qquad (2)$$

where ΔC_m is the total concentration gradient of all impermeable species, and ΔC_s is the concentration gradient of the single permeable solute under study. The term $\bar{C}_s$ is a mean value of permeable solute concentration within the membrane defined by $\bar{C}_s = \Delta C_s / \ln(C_s/C_s^i)$, where C_s and C_s^i represent permeable solute concentrations outside and inside of the cell. The first equation describes water flow in response to hydraulic and osmotic gradients across the membrane. The second is of primary interest here since the last term contains P_s, whose behavior will indicate whether or not carrier kinetics are involved for the penetrant being studied. The first term in *equation 2* is a solvent drag term. An intuitive understanding of this term is obtained by considering the sieve properties of a membrane with pores: if the penetrant is large enough, none will be dragged through the membrane as water flows through; if the penetrant is small enough, all dissolved in the water will go across the membrane with the water. The magnitude of the solvent drag term is expressed by σ, which can be thought of as the fraction of solute molecules that cannot move through the pores. Thus, if $\sigma = 1$, all solute molecules are reflected; if $\sigma = 0$, none are reflected.

These equations can be rewritten in a more useful form by relating the volume flux to changes in cell volume V, and solute flux to changes in internal concentration C_s^i:

$$\frac{dV}{dt} = 5.76\, L_p \left[\frac{0.6}{V - 0.4} - C_m - \sigma(C_s - C_s^i)\right] \qquad (3)$$

$$\frac{dC_s^i}{dt} = \left[((1 - \sigma)\bar{C}_s - C_s^i)\,\frac{dV}{dt} + 1.92 \times 10^4 P_s (C_s - C_s^i)\right] / (V - 0.3) \qquad (4)$$

Equations 3 and *4* are written in dimensionless form. The numerical constants incorporate relevant cell parameters such as membrane area together with appropriate scaling factors. The computer solutions were essential in three important ways, namely for designing experiments, for checking data processing, and for testing alternative hypotheses. An example of these solutions is obtained by simulating an experiment where the cells

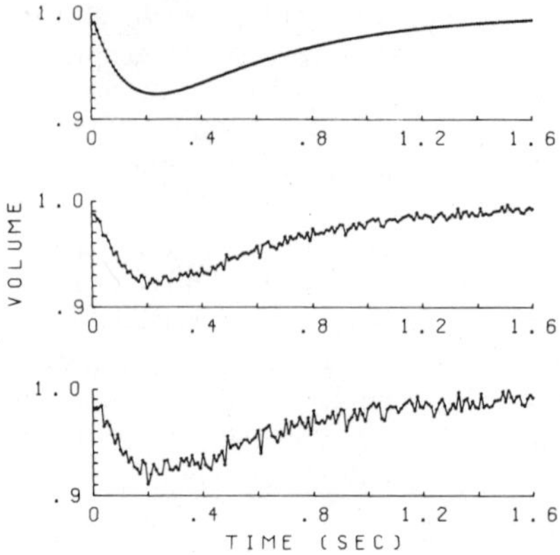

Figure 1. The curves show changes in cell volume (in arbitrary units) with time in seconds due to a step increase in solute concentration. The curves were obtained by computer simulation using *equations 3* and *4*. An arbitrary amount of random noise was superimposed on the lower two curves to simulate real data.

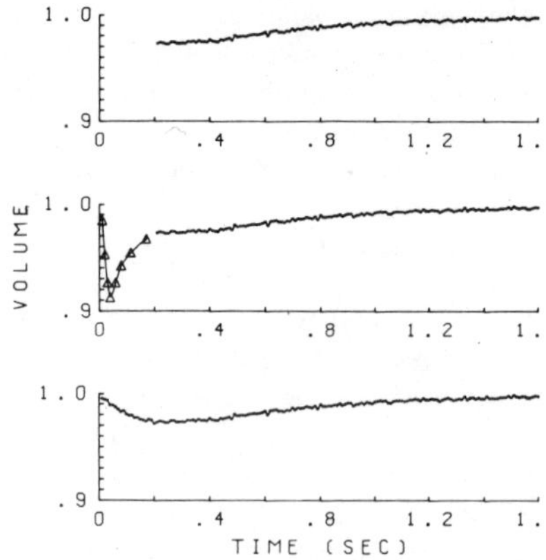

Figure 2. The upper curve shows data consisting of cell volume versus time. The first 200 msec cannot be recorded reliably and no data are given here. Addition of triangular points by hand in the middle curve shows our original (mistaken) impression of the initial portion. The lower one shows by computer simulation what the curve really looks like.

are suddenly exposed to a step increase in penetrant concentration. The top curve in Fig. 1 shows the resulting time course of the volume change. As is expected, the cells first shrink due to water flow out in response to the increased osmotic pressure on the outside, and then swell as the penetrant, in response to its concentration gradient, enters and is followed by water. The other curves show a superposition of arbitrary amounts of noise in order to simulate real data. The noise can be simulated internally on the computer from a random noise generator, or better still, it can be recorded directly from the experimental apparatus.

Our approach to the problem of parameter retrieval was to restrict experiments to small perturbations in volume and solute concentrations. This allowed *equations 3* and *4* to be linearized and to be described by the sum of two exponentials. Data processing consisted of resolving the time constants by the method of least squares on the computer. These time constants are then related to the transport parameters.

In order to accomplish this, the step increase in urea concentration must be large enough to be measured, but small enough to allow use of the linearized equations. Before the computer was used to help design experiments, we picked our step increase in urea concentration by assuming that urea moves relatively slowly compared to water. For the experiment shown in Fig. 2, for example, we expected a 10–15% change in cell volume. As shown in the top curve in Fig. 2, reliable measurements cannot be made less than 200 msec after the urea increase, and we just assumed our signal was lost here. The middle curve in Fig. 2 shows what we thought the curve should look like. Computer simulation, given by the lowest curve in Fig. 2, showed us that we were not missing the signal. It was never there. In order to elicit a 10% volume change, we needed about four times greater increase in urea concentration.

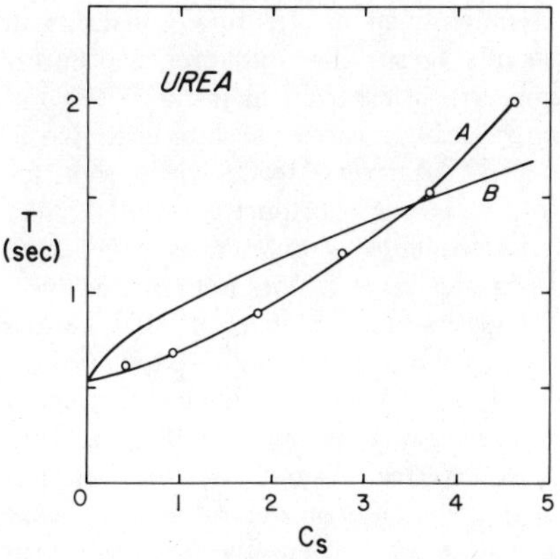

Figure 3. Our experimental data consisting of time constants in seconds and urea concentrations in normalized units are represented by circles. Curve A is the computer fit of the data for carrier kinetics ($K = 1.8$ M, $v_m = 5.2 \times 10^{-7}$ moles/(cm^2 sec)) and curve B for simple passive kinetics.

The computer was very useful in determining the accuracy of our data processing in the following way. Reasonable transport parameters were given to the computer along with experimental conditions. From these, the computer program generated a volume versus time curve upon which noise was superimposed, like those shown in Fig. 1. This "computer data" was then processed as real data to see if the original transport parameters could be recovered. We found a number of design constraints had to be imposed before we could trust the parameters measured. These constraints involved a priori knowledge of the parameters to be measured. This involved us in an iterative loop between the computer and experimental apparatus where we initially guessed the parameters, designed and performed the experiment, processed the

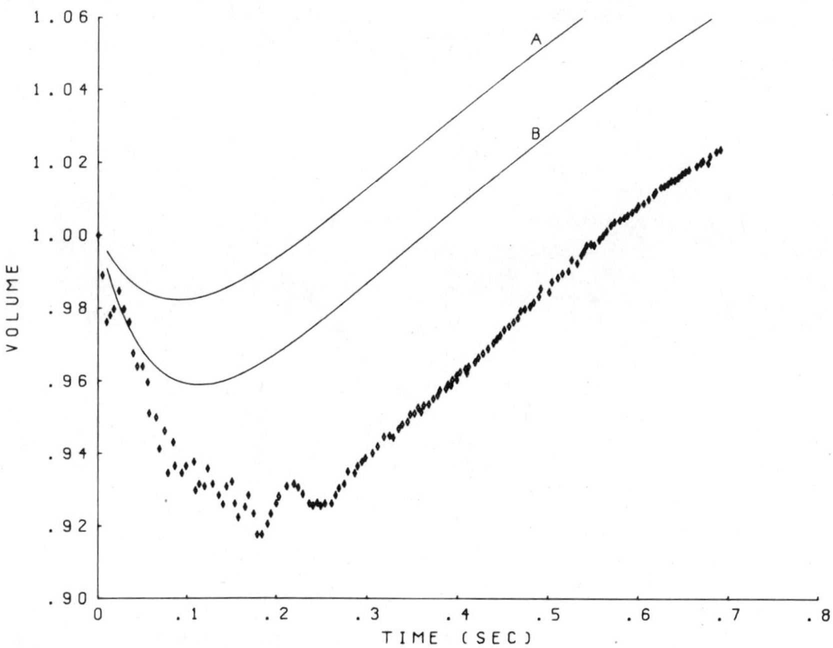

Figure 4. The data of Sha'afi et al. (10), consisting of cell volume (in arbitrary units) versus time in seconds, are given in symbols. Curve A and B are computer simulations using the experimental conditions of Sha'afi et al., their value of RTL_p (0.19 cm^4/(osM sec)) and their relationship between P_s and J_v. Curve A is for σ of 0.6 and curve B for 1.0. It can be seen that, no matter what σ is used, the computer simulations do not fit the experimental data.

data to arrive at better parameter guesses, and so on until the measured and guessed parameters converged.

Using these small perturbation procedures, we found that *equations 3* and *4* would fit the data, but as we increased the concentration of urea, the value of P_s decreased. Furthermore, the concentration dependence of P_s corresponded to predictions based on simple carrier kinetics. This was tested by replacing the term P_s in *equation 4* by the term

$$\frac{v_m K}{(C_s + K)(C_s^{\circ} + K)} \qquad (5)$$

where v_m and K are constants representing the maximal transport velocity and carrier–solute dissociation constant respectively. Figure 3 shows the experimental data given as the time constants in seconds versus the final urea concentrations in the external medium. The computer fit for a carrier system is shown to fit the data very well, while that for simple passive transport does not.

The conclusion that urea is transported via carrier kinetics does not enjoy universal agreement. Sha'afi et al. (10) assume that urea moves by simple passive diffusion but that P_s is a linear function of volume flow. Their method for evaluating P_s suffers from some equivocation, because it depends on obtaining the second derivative of an empirical curve with noise. Using a computer simulation, we used the experimental conditions of Sha'afi et al., their value for L_p, their relationship of P_s and J_v, and various values of σ. Figure 4 shows that the

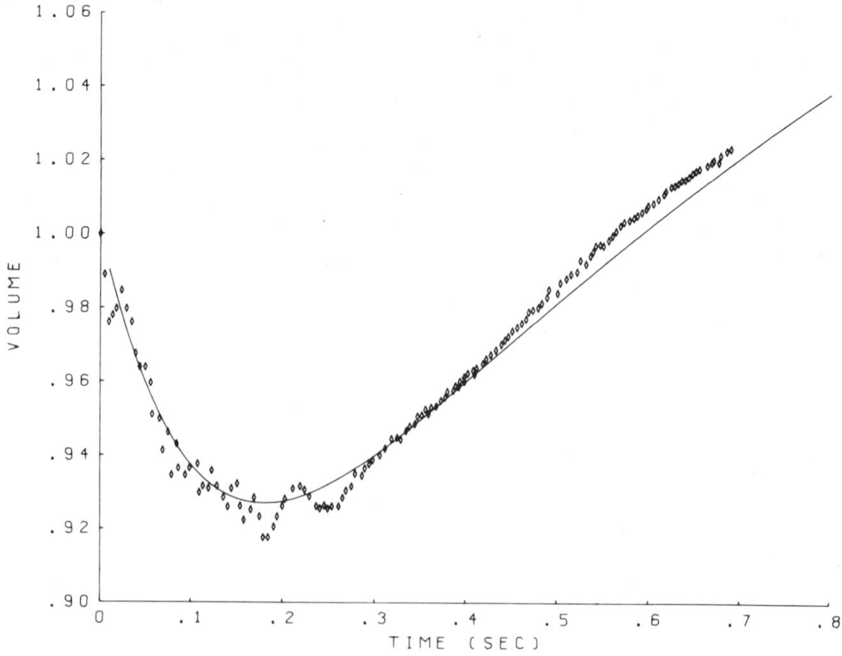

Figure 5. The data of Sha'afi et al. (10), consisting of cell volume (in arbitrary units) versus time in seconds, are given in symbols. The smooth curve was obtained by computer using the experimental conditions and value of RTL_p (0.19 cm^4/(osM sec)) of Sha'afi et al. The P_s value used was about 150% smaller than that of Sha'afi et al. at J_v of 0. For this fit, σ = 1.0, P_s = 2.3 $\times$ 10^{-4} cm/sec and RTL_p = 0.19 cm^4/(osM sec).

computer simulations do not fit their experimental data no matter what value of σ is used. However, a constant P_s, smaller by about 150% of Sha'afi et al.'s P_s at J_v of 0, fits the data quite well, as Fig. 5 shows. The data are fit best by a σ of 1.00. This in itself would suggest that urea does not penetrate through pores, but the value is not unique and will decrease if a larger value of L_p is used.

Our experimental data and computer simulations similar to those illustrated in the last few paragraphs have convinced us that the available data are incompatible with simple diffusion, but easily explained in terms of carrier kinetics. They suggest that pores are major channels for water transport but relatively ineffective for solutes.

These simulation studies were relatively straightforward, but they depended on having an easily interactive computing system. They were carried out on a PDP-7 using a simulation language which was recently written by David Auslander (1). The language has the following features:

1) It is easy to program—undergraduates with no computer experience become fully conversant with the language with only 1 hour's experience.

2) Significant problems can be solved on a mini-computer with as little as 4,000 words of memory and nothing but a teletype for input–output. Cost estimates for use with graphical outputs are about \$2/hour as compared to \$10 to \$15/hour on commercial time-sharing systems with no graphical outputs. It is feasible to put the whole system in the hands of an individual investigator.

3) The language is highly interactive. Solutions are obtained within a few seconds and parameter and even equation changes are accomplished with single instructions.

REFERENCES

1. Auslander, D. M. *J. Dynamic Systems, Measure, Control. Trans. ASME.* Series G, No. 3, 96: 261, 1974.
2. Cole, K. S. *Membrane, Ions and Impulses.* Berkeley: Univ. of California Press, 1968.
3. Cole, K. S., R. Guttman and F. Bezanilla. *Proc. Natl. Acad. Sci. U.S.* 65: 884, 1970.
4. Diamond, J. M., and E. M. Wright. *Ann. Rev. Physiol.* 31: 581, 1969.
5. Hodgkin, A. L., and A. F. Huxley. *J. Physiol. London* 117: 500, 1952.
6. Kedem, O., and A. Katchalsky. *Biochim. Biophys. Acta.* 27: 229, 1958.
7. Levitt, D. G. *Biophys. J.* 13: 186, 1973.
8. Macey, R. I., D. M. Karan and R. E. L. Farmer. *Biomembranes* 3: 331, 1972.
9. Paganelli, C. V., and A. K. Solomon. *J. Gen. Physiol.* 41: 259, 1957.
10. Sha'afi, R. I., G. T. Rich, D. C. Mikulecky and A. K. Solomon. *J. Gen. Physiol.* 55: 427, 1970.

Cardiac surgical intensive care computer system

LOUIS C. SHEPPARD AND JOHN W. KIRKLIN
University of Alabama in Birmingham, Alabama 35294

Optimal care of critically-ill patients during the crucial hours following open heart surgery requires experienced, well trained, and highly motivated registered nurses. The limited availability of nurses in our geographic area with these qualifications is the primary reason for our employing computer-based postoperative care techniques in the Cardiac Surgical Intensive Care Unit of the University of Alabama Hospital, Birmingham, Alabama. We have automated many of the routine tasks in the CICU and by this means we are able to maintain the required high standard for quality of care despite constantly increasing patient loads and chronic CICU nursing staff shortages (5).

BRIEF DESCRIPTION OF CICU SYSTEM

Connected to the process control computer system are medical electronics and instrumentation for monitoring heart rate, intraarterial pressure, right and left atrial pressures, rectal temperature, chest tube drainage, and urine output (4). Clinical measurements are acquired automatically by the system at 2-minute intervals and displayed at the bedside. Periodically, current measurements are stored by the computer for retrieval to prepare the printed record for the hospital chart and to permit the bedside review of trends in patient status (5). Part of the real work of patient care is the automatic control of left atrial pressure by blood infusion. This unique involvement of the computer in the actual delivery of therapy in a closed-loop, feedback mode is based on sound physiological principles (6) and utilizes computer-controlled infusion subsystems designed to function safely and reliably in a clinical environment. Control of the computer from the bedside is achieved by means of video display terminals equipped with small keyboards that are used to direct the computer to carry out certain predetermined tasks, select options displayed by the computer on the video monitors, and enter numeric data such as arterial blood gas measurements (5). Eight of the 12 beds in the CICU are instrumented as described and the automated system is fully integrated into the daily operation of the unit.

USE OF THE SYSTEM

On the day preceding surgery a single data card is punched for each patient scheduled for cardiac surgery employing cardiopulmonary bypass. In the card is punched the patient's name, hospital number, age, date of operation, sex, height, weight, and the code which specifies the antibiotics to be administered. From this information the standard preoperative and postoperative orders are prepared; body surface area is calculated; the mean new intravenous dose of digoxin and the amount of intravenous fluid to be administered are calculated, as well as the dosage of the antibiotics and the instructions for their administration (3). All of the information in the patient data cards as well as the calculated results are tabulated and printed on the line printer in a format that is convenient for the nurses.

At the end of the operation the patient is taken directly to the Cardiac Surgical Intensive Care Unit. The CICU technician in the unit has previously prepared the computer for the patient, cleaned, sterilized, and calibrated the pressure transducers as well as installed the chest drainage jug and urine collection bag in the protective frame enclosure. Immediately upon admission to the unit, the patient's electrocardiograph leads are connected to the appropriate electronics, and the computer is directed to begin automatic measurements. Next, measurement of chest drainage is initiated. The interconnecting tubing from intravascular cardiac catheters inserted at surgery are coupled to the proper pressure transducers, the zero references are adjusted, and the measurements are automatically obtained and analyzed by the computer. The blood infusion pump is readied with the unit of blood. Automatic infusion is specified, and the pressure limit entered. When the urinary catheter has been inserted by the nurse or physician, automatic measurement of urine output is initiated. Insertion of the rectal thermistor probe and activation of this measurement is the final step (4). This process requires a total of 10–15 min.

The patients are transferred out of the CICU to their hospital rooms on the day following surgery, unless further respiratory care is required or antiarrhythmic, vasoactive or inotropic drugs are still being administered. Because of the automated system, continuation of blood infusion beyond 7 AM is unnecessary in nearly all cases. After extubation, withdrawal of intracardiac and intraarterial cannulae, and removal of chest drainage tubes, transfer can be considered. In most cases this stepwise procedure has been completed and the patient transferred by noon of the day after operation.

TABLE 1. Response of left atrial pressure and stroke index to blood infusion as a function of initial pressure

	Blood volume infused, ml	Left atrial pressure, mm Hg	Stroke index, ml/m²
Group I[a]	0	7.1	26.6
	216	10.6	31.4
Group II	0	15.6	27.4
	129	19.7	24.7

[a] *Group I,* left atrial pressure less than 12 mm Hg; *Group II,* left atrial pressure greater than 12 mm Hg.

AUTOMATIC CONTROL OF THE INFUSION OF BLOOD

Convenient methods for reliable, repetitive measurement of stroke volume (in milliliters) in the clinical environment do not yet exist. Consequently we have based the closed-loop control of blood infusion on left atrial pressure (6), a measurement which is considered to be one of the determinants of stroke volume (and hence cardiac index, liter/minute/square meter).

The data (1) shown in Table 1 are from two groups of patients (nine each) which illustrate the effect of rapid infusion of a large volume of blood on left atrial pressure and stroke index. In group I, those patients with low or normal beginning left atrial pressure, the blood infusion increased pressure and stroke index. The patients in group II, with left atrial pressures exceeding 12 mm Hg, did not respond favorably to the rapid infusion. Pressures increased as in the first group but with a smaller volume of infusion. Furthermore the stroke index (and cardiac index) fell. Based on these and other data, patients with low cardiac index and high left atrial pressure do not receive blood while these conditions exist.

A composite diagram, Fig. 1, has been prepared from these data to illustrate the underlying physiological concepts employed in the computer-programmed blood infusion control.

One of the factors which determines stroke volume is end diastolic volume, the volume of blood the ventricle contains just prior to contraction (EDV, upper left quadrant). End diastolic pressure (EDP), preload, is useful in following possible changes in end diastolic volume (upper right quadrant). We have no technique for clinically measuring end diastolic volume but have found left atrial pressure to be closely correlated with end diastolic pressure and feasible to measure (lower right quadrant). So left atrial pressure is the variable controlled in the automatic blood infusion (lower left quadrant) since it ultimately correlates with stroke index (6).

The dashed line illustrates the interrelationships of the variables by tracing the measured effect of blood infusion on left atrial pressure and stroke index as well as the hypothesized effect on end diastolic pressure and end diastolic volume.

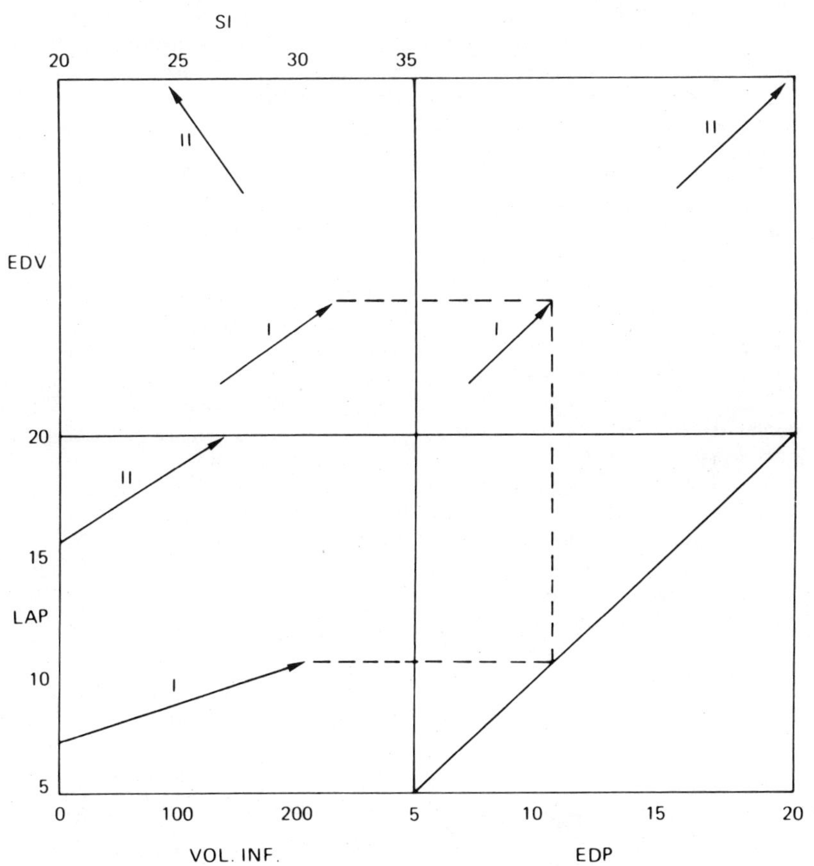

Figure 1. Physiological concepts employed in computer-controlled blood infusion. See text for explanation of figure.

It was 6.5 years ago that we first programmed our computer to take an active part in the actual control of therapy by automatically infusing blood if certain measurements did not exceed specified limits (6). The essential elements of the computer-controlled blood infusion pump are the control unit, a drip sensing assembly, and a roller pump driven by a stepping motor (4, 5).

The infusion of blood is automatically controlled by the computer system in a closed-loop feedback mode based on left atrial pressure which is related to stroke volume and hence cardiac output (CO). A left atrial pressure limit is either set automatically by the computer or specified by the physician to represent that level which is associated with optimal stroke volume. If the infusion of blood does not result in the left atrial pressure measurement reaching this limit, the total blood infused (B) is then limited to be some multiple of the measured drainage (D) in order to avoid over-infusion. If the cardiac output has been measured by the indicator dilution method, the limiting values of left atrial pressure and the total blood infused-to-drainage ratio (B/D) can be set automatically by the computer. Otherwise the physician can specify the limits individually. A 20-ml dose of blood is automatically infused each 2 min if the left atrial pressure measurement does not exceed its limit. If left atrial pressure is above its limit or if the current B/D exceeds the allowable ratio, no blood is infused. In the first few hours after operation the patient's measured chest tube drainage can be sufficiently low to limit the volume of blood infused. Therefore the automatic infusion rules are based solely on left atrial pressure until the total volume infused per square meter of body surface area exceeds 250 ml/m^2.

IMPAIRED CARDIAC PERFORMANCE

The analysis, decision-making, and treatment of the cardiovascular subsystem during the early postoperative period are based upon assessing the blood flow (cardiac output or cardiac index), its adequacy, and the likelihood that intervention will result in the desired effect (reserves). If the function of the heart is observed preoperatively to be impaired, low cardiac output is likely to be observed postoperatively as well. Since cardiac output is the product of heart rate and stroke volume, the first step is to achieve optimal heart rate by pacing if the intrinsic rate is not over 90 per min.

The computer-programmed decision tables (5) for analysis and decision-making regarding selection of interventions were derived from clinical research studies of three factors that affect stroke volume: *preload*, the initial stretch of the ventricular muscle fibers just preceding ejection (1); *contractility*, related to how vigorously the ventricle contracts (snap); and *afterload*, the external circuit load that resists the shortening of the muscle fibers when the ventricle contracts (impedance) (2).

The variables that we use to describe the decision domains relating to the factors are left atrial pressure (LA), mean arterial pressure (MAP), and cardiac index (CI):

LA $\leq$ 14, CI $<$ 3.0
increase preload with blood infusion to raise stroke volume

LA $>$ 14, CI $<$ 1.8
increase contractility with epinephrine/isoproterenol to raise stroke volume

LA $>$ 18, CI $<$ 2.2
MAP $>$ 100
decrease afterload with Trimethaphan to raise stroke volume

Knowledge of cardiac index enables the physician to interrogate the computer regarding the recommended intervention(s).

RESULTS

In the past 7 years the automated care system has been used in the postoperative observation and treatment of 4,150 patients. This system has enabled us to supplement our relatively small staff of registered nurses with paramedical personnel, CICU technicians. Since routine, repetitive tasks that are well defined have been relegated to the system, the nurses are able to devote a higher percentage of their time to direct patient care. Consequently, we are able to serve more patients with the existing nursing staff than would be possible without the system and the CICU technicians. Compared to the usual manual methods of measurement, the automated measurements are more accurate, more reliable, and more consistent. Furthermore, missing data due to oversight, distraction, or work load have been eliminated as a problem. Trends in the patient's condition can be assessed more accurately from the automated measurements. The past data can be retrieved more quickly by way of the bedside terminal than by search through the patient's chart. Physicians trust the computer-generated data more than the data transmitted by people. The explicit nature of the system reduces the tendency to apply judgment unnecessarily in those situations that have well-defined decision-making rules. Since the logic of experienced surgeons is embodied in the standardized operation of the system the span of control of senior surgeons is increased. The stability of the man–machine system enhances the continuity of care delivered to the critically ill patient. The care of the individual patient is not compromised by the need for his nurse to temporarily assist with a crisis elsewhere in the unit. The patient's postoperative course in the crucial early hours following surgery is more stable because the computer-controlled blood infusion is more objective, less erratic, and more intense than manual methods.

REFERENCES

1. Kouchoukos, N. T., J. W. Kirklin, L. C. Sheppard and P. A. Roe. Effect of left atrial pressure by blood infusion on stroke volume early after cardiac operations. *Surg. Forum* 22: 126, 1971.
2. Kouchoukos, N. T., L. C. Sheppard and J. W. Kirklin. Effect of alterations in arterial pressure on cardiac performance early after open intracardiac operations. *J. Thorac. Cardiov. Surg.* 64: 563, 1972.
3. Kouchoukos, N. T., L. C. Sheppard and J. W. Kirklin. Automated patient care following cardiac surgery. *Cardiac Surg.* 3: 110, 1971.
4. Sheppard, L. C., J. W. Kirklin and N. T. Kouchoukos. Computer-controlled interventions for the acutely ill patient. *Comput. Biomed. Res.* 4: 135, 1974.
5. Sheppard, L. C., N. T. Kouchoukos and J. W. Kirklin. The digital computer in surgical intensive care automation. *Computer* 6: 29, 1973.
6. Sheppard, L. C., N. T. Kouchoukos, M. A. Kurtts and J. W. Kirklin. Automated treatment of critically ill patients following operation. *Ann. Surg.* 168: 596, 1968.

Session I—Computer applications I
Rapporteur's summary

HELEN HOFER GEE
National Institutes of Health
Bethesda, Maryland 20014

This meeting marks a milestone for Biomedical Computing. It announces a readiness to be recognized as an entity in the bioscience community and perhaps in the political arena as well. In 1965, when Bruce Waxman left the Biomedical Computer Study Section at the NIH, he said to the committee, in effect, "It's been great, fellows, starting to fund biomedical computer research and facilities, and getting a new field going, but in all honesty I can't think of a single major contribution biomedical computing has made to the advancement of biomedical research." Just 9 years later, at this conference, it has taken the better part of a week for our speakers to skim lightly over a wide range of scientific advances in which the computer has played an absolutely critical role. From another perspective, it may be observed that NIH now supports research in which the computer plays a major role at a rate close to $100 million per year. Although this figure represents the total costs of the identified research, not just the computer costs, it commands attention. It emphasizes a point made by the substance of most of the papers presented at this conference—that the computer is no mere alternative tool in today's bioscience effort, it has become an integral component of the scientific process itself in many areas of investigation. The ramifications of these observations are many, but they are not my privilege to explore here. I must turn to my assigned role which is to summarize the first session of this conference.

Don Lindberg set the stage for the conference outlining three objectives: to assess the contribution of the computer to the Life Sciences; to identify areas that might afford leverage in future development; and to explore the problem of survival of computer-based research, with concern especially for how hardware and funding policies influence the character of that survival.

The first session, chaired by Julia Apter, addressed two of these objectives as speakers described a variety of applications of computer technology in biomedical research and medical service activities. In addition to outlining successful achievements, some speakers and most of the discussion periods also addressed problems that must be dealt with if further advances are to build on present accomplishment. One might add parenthetically that in the absence of focused Federal concern with the problems of biomedical computing, finding opportunities to mount concerted attacks on such problems may itself be critical.

Dr. Margaret Dayhoff began the substantive program with a description of her impressive *Atlas of Protein Sequencing and Structure*. The *Atlas* is a testament to the contributions of computing to one broad area of scientific investigation. A particularly exciting aspect of this work is the development of the field of molecular evolution which promises to increase manyfold the knowledge of evolution that has heretofore been the almost exclusive domain of the geneticist.

Dr. John Melski, speaking for Dr. Howard Bleich, demonstrated how string-manipulating computer languages and interactive terminals have permitted the development of a computer-based physician consulting service. Dr. Bleich's system can enhance significantly the quality of medical services most physicians are able to provide in the areas of electrolytes and acid–base balance. There is no question that it constitutes an important positive contribution to the problem of extending the availability and quality of medical care. But the development of other similar systems in other areas of clinical service is not that clearly indicated, as was brought out in the discussion that followed the paper. One problem is that tracking diagnoses and treatment alternatives is seldom so well delineated as in the areas of electrolytes and acid–base. How many similarly well delineated areas exist that would yield to such an approach? Another kind of problem is concerned with the question of when and under what circumstances such systems should be changed and updated. Costs and problems of communication must be balanced against the significance of prospective changes. In areas where the knowledge base is controversial, whose opinion is to be allowed to guide other physicians? How can idiosyncrasies be safely introduced into such systems? In still another vein, what is it going to take to get third party insurance support to pay for such systems? And what is the Federal role in their development and support? None of these questions have been decided nor have they been explored in any depth.

Dr. Carol Newton laid it on the line with respect to the status, the problems, and the prospects in the interface between computer science and the field of biomathematics, including biostatistics. She faced up to the truth that no statistical system is or probably ever will be all things to all the analytical needs of biomedical researchers, and she delivered a challenge for the development of computer-based statistical aids to research planning as well as to data analysis. In an extra session Dr. Newton outlined exciting new prospects for data handling through interactive graphics. Computer aids to biomathematics/biostatistics have probably resulted in more assistance to more biomedical researchers than any other computer application. The area also generates some sticky problems. Consider the many facets of the problem of accessing the "right" computer system for any given bioscience project. Individual investigators may or may not have the technical knowledge needed to access a particular system, and may or may not be able to find or to buy good consulting help. Many investigators have difficulty finding funds to pay for computer usage, and sometimes, though funds are available, investigators are unable to gain acceptance as a user on a particular facility. There are problems of geography which networking and time-sharing can solve only partially. Recently a new problem of campus politics has developed in which one faction would have decisions about access to computer systems based on the financial needs of the local computing facility. There exist too a whole series of grave educational problems, varying with time, place, individuals and institutions, from resistance on the part of students (and some faculty) to learning about the theoretical and analytical bases of biology, to institutional myopia in failing to recognize the importance of the problem or to provide needed educational opportunities. Sudden changes in and accompanying uncertainty about Federal policy concerning the support of training contribute to these difficulties. The problems are too broad and too complex to address here. They are difficult for both academia and government, and warrant the active concern of the entire research community.

Dr. Robert Macey introduced to the conference the exciting world of model development describing an application to the area of membrane transport. The discussion of his paper exposed the problem the modeler has of gaining acceptance of his particular approach, but mainly it provided a taste of the intellectual excitement that modeling generates among both doers and observers.

Finally, Dr. Louis Sheppard presented us with another major achievement in the clinical application of the computer, this time to the delivery of cardiac intensive care. The system he described provides extensive data acquisition, analysis, display and storage capabilities for the use of the responsible medical staff. It includes as well a system for automatic control of the infusion of blood in a closed-loop, feedback mode, and provision for the detection of critical changes in chest tube drainage patterns. Again, admiration of technical and scientific accomplishment was accompanied by recognition of problems and questions. One participant raised a familiar sort of question about the efficacy of continuing the future development of the University of Alabama cardiac intensive care monitoring system on "obsolete 1800 computers." Another responded most efficaciously pointing out that it is too bad that people lose sight of the fact that a system on which a program is developed will always be able to do the job; change is not indicated until the system ceases to be appropriate.

In another vein, the question opens up a wide range of problems that can be summarized as problems in the diffusion of computer-based technology. At this juncture biomedical computing joins all the rest of biomedicine. The problems of diffusion of advances in health research, including advances in biomedical computing, are commanding increasing attention, particularly as prospects increase for the expansion of health care delivery through national insurance coverage and wider distribution of health care personnel. Little useful documentation is available about these problems, which require not only solutions to specific technical or administrative problems, but the development of knowledge and understanding of basic social and behavioral principles.

This summary attends more to problems than to review of the substance of the many important advances in the biosciences described by the speakers, but the very emergence of problems attests to the lively vigor of the biomedical computing scene.

Session II

Computer applications II

Introductory remarks

EDWARD A. FEIGENBAUM, *chairperson*
Stanford University, Stanford, California 94305

This session on Computer Applications combines papers on three disparate topics: advanced graphical display techniques in the service of protein chemistry and biology; applications of Artificial Intelligence (AI) research; and biological laboratory automation. My introductory remarks will concern themselves only with AI Research.[1]

Artificial Intelligence research is that part of computer science that is concerned with the symbol-manipulation processes that produce intelligent action. By "intelligent action" is meant an act or decision that is goal-oriented, arrived at by an understandable chain of symbolic analysis and reasoning steps, one in which knowledge of the world informs and guides the reasoning.

The potential uses of computers by people to accomplish tasks can be "one-dimensionalized" into a spectrum representing the nature of instruction that must be given the computer to do its job. Call it the "WHAT-TO-HOW" spectrum. At one extreme of the spectrum, the user supplies his intelligence to instruct the machine with precision exactly HOW to do his job, step-by-step. Progress in computer science can be seen as steps away from the extreme HOW point on the spectrum: the familiar panoply of assembly languages, subroutine libraries, compilers, extensible languages, etc. At the other extreme of the spectrum is the user with his real problem (WHAT he wishes the computer, as his instrument, to do for him). He aspires to communicate WHAT he wants done in a language that is comfortable to him (perhaps English); via communication modes that are convenient for him (including, perhaps, speech or pictures); with some generality, some abstractness, perhaps some vagueness, imprecision, even error; without having to lay out in detail all necessary subgoals for adequate performance with reasonable assurance that he is addressing an intelligent agent that is using knowledge of his world to understand his intent, to fill in his vagueness, to make specific his abstractions, to correct his errors, to discover appropriate subgoals, and ultimately to translate WHAT he really wants done into processing steps that define HOW it shall be done by a real computer. The research activity aimed at creating computer programs that act as "intelligent agents" near the WHAT end of the WHAT-TO-HOW spectrum can be viewed as the long-range goal of AI research. Historically, AI research has always been the primary vehicle for progress toward this end, even though science as a whole is largely

[1] In the remarks that follow, I wish to acknowledge my debt to Allen Newell of Carnegie-Mellon University not only for numerous stimulating discussions about artificial intelligence (AI) over the last two decades, but also for the use, with his permission, of some choice unpublished sentences of his.

unaware of the role, the goals, and the progress made.

The root concepts of AI as a science are *1*) the conception of the digital computer as a symbol-processing device (rather than as merely a number calculator); and *2*) the conception that all intelligent activity can be precisely described as symbol manipulation. (The latter is the fundamental working hypothesis of the AI field, but is controversial outside of the field.) The first inference to be made therefrom is that the symbol manipulations which constitute intelligent activity can be modeled in the medium of the symbol-processing capabilities of the digital computer.

This intellectual advance—which gives realization in a physical system, the digital computer, to the complex symbolic processes of intelligent action and decision, with detailed case studies of how the realization can be accomplished, and with bodies of methods and techniques for creating new demonstrations—ranks as one of the great intellectual achievements of science, allowing us finally to understand how a physical system can also embody mind. The fact that large segments of the intellectual community do not yet understand that this advance has been made does not change its truth or its fundamental nature.

HIGH PERFORMANCE PROGRAMS THAT PERFORM AT NEAR-HUMAN LEVEL IN SPECIALIZED AREAS

As the AI research matured to the point where the practioners felt comfortable with their tools, and adventuresome in their use; as the need to explore the varieties of problems posed by the real world was more keenly felt; and as the concern with knowledge-driven programs (to be discussed later) intensified, specific projects arose which aimed at and achieved levels of problem solving performance that equaled, and in some cases exceeded, the best human performance in the tasks being studied. An example of such a program is the heuristic DENDRAL program, developed by our interdisciplinary group at Stanford University. It solves the scientific induction problem of analyzing the mass spectrum of an organic molecule to produce a hypothesis about the molecule's total structure. This is a serious and difficult problem in a relatively new area of analytical chemistry. The program's performance has been generally very competent and in "world's champion" class for certain specialized families of molecules. Similar levels of successful performance have been achieved by some of the MATHLAB programs at MIT, that assist scientists in doing symbolic integration in calculus, and are virtually unexcelled. In medical problem solving tasks, applications are beginning to be developed. One of these is the program under development at Rutgers concerning the eye disease glaucoma, that will be discussed by Professor Amarel in his paper. Another is MYCIN program (being developed at the Stanford University Medical School) for advising physicians about antibiotic therapy for treating infectious diseases. The NIH has recently established a national computer facility (SUMEX) for application of AI to problems in biology and medicine—a facility located at the Stanford University Medical School but accessible by the national community of interest over data communication networks.

KNOWLEDGE-BASED SYSTEMS

This term is intended to connote, in familiar terms, something like: "What is the meaning of . . ." or "How is that to be understood . . ." or "What knowledge about the world must be brought to bear to solve the particular problem that has just come up?" The research deals with the problem of extracting the meaning of: utterances in English; spoken versions of these; visual scenes; and other real-world symbolic and signal data. It aims toward the computer understanding of these as evidenced by the computer's subsequent linguistic, decision-making, question-answering, or motor behavior.

Thus for example, we will know that our "intelligent agent" understood the meaning of the English command we spoke to it if: *a*) the command was in itself ambiguous; *b*) but was not ambiguous in context; and *c*) the agent performed under the appropriate interpretation and ignored the interpretation that was irrelevant in context.

One paradigm for work in this area of knowledge-based systems is, very generally sketched, as follows:

> *a*) a situation is to be described or understood; a signal input is to be interpreted; or a decision in a problem-solution path is to be made. *Examples:* A speech signal is received and the question is, "What was said?" The TV camera system sends a quarter-million bits to the computer and the question is, "What is out there on that table and in what configuration?" The molecule structure-generator must choose a chemical functional group for the "active center" of the molecular structure it is trying to hypothesize, and the question is, "What does the mass spectrum indicate is the 'best guess'?"
>
> *b*) Specialized collections of facts about the various particular task domains, suitably represented in the computer memory (call these Experts) can recognize situations, analyze situations, and make decisions or take actions within the domain of their specialized knowledge. *Examples:* In the Carnegie-Mellon University Hear-say speech understanding system, currently the Experts that contribute to the current best hypothesis are an acoustic–phonetic Expert, a grammar Expert, and a chess Expert (since chess playing is the semantic domain of discourse). In heuristic DENDRAL, the Experts are those that know about stability of organic molecules in general, mass spectrometer fragmentation processes in particular, nuclear magnetic resonance phenomena, and so on.

For each of the sources of knowledge that can be delineated, schemes must be created for bringing that knowledge to bear at some place in the ongoing analysis or understanding process. The view is held that programs should take advantage of a wide range of knowledge, creating islands of certainty as targets of opportunity arise, and using these as anchors for further uncertainty reduction. It is an expectation that always some different aspect provides the toehold for making headway—that is, that unless a rather large amount of knowledge is available and ready for application, this paradigmatic scheme will not work at all.

One of the most interesting and successful of the new knowledge-based systems employing the paradigm just sketched is the Hearsay speech understanding computer system being developed by Professor Reddy and his group at Carnegie-Mellon University, and described in Reddy's presentation to follow.

To summarize: In AI research, there are foci upon the encoding of knowledge about the world in symbolic expressions so that this knowledge can be manipulated by programs; and the retrieval of these symbolic expressions, as appropriate, in response to demands of various tasks. AI has sometimes been called "applied epistomology" or "knowledge engineering."

The AI field has come increasingly to view as its main line of endeavor: knowledge representation and use, and an exploration of understanding (how symbols inside a computer, which are in themselves essentially abstract and contentless, come to acquire a meaning).

The impact of the "knowledge engineering" research and development endeavors of the AI scientists can and ultimately will have a major impact on the organization of biomedical knowledge; on its use in the service of intelligent computer assistance to practitioners, researchers, and educators; and on the transmission of this knowledge in medical education.

Interactive three-dimensional computer graphics in molecular biology[1]

ROBERT LANGRIDGE
Princeton University
Princeton, New Jersey 08540

Interactive three-dimensional computer graphics was first used in 1964 to study biological macromolecules, particularly nucleic acids (3) and proteins (1). A good general description of this early work is given in (4). The hardware was the then unique Project MAC Electronic Systems Laboratory display at MIT. It was some time before comparable equipment became available elsewhere.

Similar developments have since been reported in this field using commercially available equipment and other special purpose noncommercial equipment.

At Princeton in 1969 the Computer Graphics Laboratory was initiated.[2] The basic configuration is an Evans and Sutherland LDS-1 interfaced to a Digital Equipment Corporation PDP-10. The capabilities of this greatly improved computer graphics system in studying macromolecular structures are illustrated in (2).

Detailed written descriptions of equipment become obsolete so rapidly that I will confine myself to a broad view of the advantages and disadvantages as well as the desirable features of an interactive three-dimensional computer graphics system, illustrating these features by reference to our system.[3]

The sheer size and rapidly increasing number of known structures of large biological molecules makes three-dimensional interactive graphics an indispensible tool in studying these macromolecules. Interactive graphics has advantages for visualizing not only static studies of known structures, but also for their dynamic conformation changes and interactions with other molecules.

Unlike real models, the graphics system can retain the precise coordinates of the structure seen, the coordinates of modified structures are available at once, and the data and display of previous structures can be stored and recovered easily. A good display program also allows the user to select interactively those regions of interest and, misusing an old metaphor, to study individual trees, or small groves, rather than the entire incomprehensible forest.

For the convenient study of large molecules necessary features include the ability to draw several thousand lines flicker-free and some form of hardware rotation —either digital or analog. The latter is very important, as smooth rotation of

---

[1] Supported by NIH Research Grants RR-578 and GM-16539; NSF GB-28021; and by Princeton University.

[2] The Princeton University Computer Graphics Laboratory was established by a grant to the author in 1970 from the Division of Research Resources of the National Institutes of Health (RR-578), and has also been funded in part by Princeton University.

[3] Required reading for anyone interested in interactive computer graphics is the excellent text by Newman and Sproull (5).

three-dimensional objects is one of the most important elements in making use of the display seem "natural" to persons used to handling "real" molecular models. We have written a variety of programs of general interest for studying proteins and nucleic acids.

CAAPS

As an example of a thoroughly tested and well documented program I will briefly describe some features of CAAPS (Computer Aided Analysis of Protein Structure). The aim of this system is to enable a computer-naive individual to study any part of any of a large number of protein structures (we presently have 21 protein coordinate files available on disk) in any detail. It was first used in early 1972 and has been updated since then in interaction with faculty, research staff and students. The description below is a condensed version of the user manual which is available on request, and which forms the basis for a more detailed report.

Creating and deleting pictures

The image on the screen is the combination of one or more CAAPS "pictures." Up to *nine* separate pictures may be defined at the same time. ("Defining the picture" always means defining, via the PICTR command, the parts of the molecule to be displayed in that picture.) Why this should be desirable will become clear in the descriptions of some CAAPS commands and in the examples.

Command language

Command: PICTR *n*

PICTR *n* begins the process of defining the parts of the protein to be displayed in a picture. "*n*" should be an integer from 1 through 9, and will become the "picture number." After executing this command, CAAPS will type a hyphen to signify that it is waiting for a picture definition command (PDC). Up to 15 PDCs may be typed to actually define the picture. After the last PDC has been typed, the user responds to the next hyphen with a carriage return, and the picture will be constructed and displayed.

Picture definition commands

Basic PDC format

A picture definition command (PDC) specifies the part of the molecule to be displayed, and indicates how the part is to be drawn on the display. A PDC consists of three parts. The first part is a single letter which indicates which portions (backbone; side chain, etc.) of the specified residues are to be displayed. The second part consists of one or two residue numbers which specify the range of residues to be displayed. The third part, which is optional, gives special drawing instructions.

PDC: <letter> <residue 1>,
<residue 2>, <drawing code>

For the first part of the PDC, the following letters are permissible:

- B specifies that only the backbone is to be drawn.
- S specifies that only the side chains are to be drawn.
- A specifies that both backbones and side chains (i.e., the entire residues) are to be drawn.
- C specifies that only lines connecting the CA atoms of adjacent residues are to be drawn. (The resulting picture, while not a true representation of connectivity, is useful for examining gross structure, and is usually the first used in studying a new molecule.)

In the second part, <residue 1> and <residue 2> are integers which specify that all residues with numbers from <residue 1> through <residue 2> are to be displayed. (That is, the parts of them specified by the first letter are displayed.)

The third part of the PDC may be an integer from 0 through 3. It has the following meanings:

0—Draw the picture with solid-line bonds, variable brightness.

1—Draw the picture with solid-line bonds, fixed bright.

2—Draw the picture with dashed bonds, variable brightness.

3—Draw the picture with dashed bonds, fixed bright.

If the argument is omitted, its default value is 0. Solid-line and dashed bonds are self-explanatory. "Variable brightness" means that this part of the picture can be varied in brightness by the CONTRAST knob from full-bright to invisible at any time. "Fixed-bright" lines are always full-bright. The brightness options, used intelligently, give rise to some very useful effects, for example, enabling one to study an active site in detail while keeping a "ghost" of the CA connections to give the general context.

Two other types of PDCs are available to enable the user to specify residues by their spatial location and by their type.

Spatial specification of residues: neighbors

PDC: <letter> 0, <residue number>, <code>

This PDC displays all the "neighbors" of residue <residue number>. <letter> and <code> have the same meanings as in the basic PDC.

Residue m is a "neighbor" of residue n. There is some atom in n which is within r angstroms of some atom in m. The value of r in CAAPS is the value of the variable RADIUS, initially equal to 7.0, which can be altered by the RADIUS command.

When this PDC is typed, CAAPS will type out, in basic PDC format, the neighbors it has found.

Command: RADIUS <r>

RADIUS is used to change the value of the variable RADIUS in CAAPS. <r> is a floating point number which becomes the new value of RADIUS. All subsequent references to residue "neighbors" make use of this new value.

Type specification of residues: residue classes

CAAPS permits definition of residue "classes," which may consist of all the residues of one or more amino acid types. CAAPS has predefined in it the classes corresponding to the residues of each amino acid type. The class names of these classes are just the three-letter amino acid abbreviations.

To display these classes:

PDC: <letter> −1, <residue class name>, <code>

Here, <residue class name> is a three-letter name of a residue class. <letter> and <code> have the normal meanings for PDCs.

Definition of residue classes

Command: CLASS

The CLASS command causes CAAPS to enter a special routine for defining residue classes. First, a class name is specified. This is a three-letter name not already being used as a class name. Then CAAPS requires a list of amino acid types for inclusion in the class; this is a list of three-letter amino acid abbreviations, separated by carriage returns. A blank line terminates the class definition and causes a return to command mode. The new class is then available for use in the residue class PDC.

Pointing to Residues

Command: POINT n, p

POINT causes a "pointer" to be created, pointing to residue n. The pointer consists of a label appearing in the lower left hand corner of the display and a dashed line from the label to the CA atom of residue n in the display. The label consists of the residue number, n, and the three-letter amino acid name of the residue. The dashed line moves with the display as it is translated and rotated.

The argument *p* is the number of the picture with which the pointer will be associated.

The incremental pointer

The last pointer created via the POINT command is termed the *incremental pointer*. Through use of switches the pointer may be "stepped through" the molecule. Turning on one switch causes the pointer repeatedly to increment by one the residue number of the residue to which it points. An adjacent switch causes decrementing of the residue. Another switch controls the speed of the incrementing or decrementing.

Labeling residues

Command: TAGS *n1, n2*

TAGS puts a small label beside the residue labeled. A label of TAGS consists of the residue number and the single letter abbreviation of the amino acid name.

TAGS enables the user to label many residues with one command. In general, TAGS labels the residues on the screen with numbers from *n1* to *n2* inclusive. However, because large amounts of text degrade the display quality, not all the residues on the display between *n1* and *n2* will actually be labeled. The general rule is only every seventh residue in a consecutive sequence of residues will be labeled. This is enough to locate residues of interest, in conjunction with the point command.

Labeling atoms

Command: LABEL *n1, n2*

LABEL is the most detailed labeling command in CAAPS. It causes individual atom labels to be added to *all* the atoms of *all* the residues on the screen with residue numbers from *n1* through *n2*. It is mainly useful in identifying specific atoms when calculating nonbonded contacts.

Neighbor variations in TAGS and LABEL commands

A variation on the TAGS and LABEL commands implements the neighbor concept. In either command, by setting the first argument equal to zero and the second argument to a residue number *n*, the effect is to tag or label all residues that are neighbors of residue *n*. (e. g., TAGS 0, <residue number>)

Structure modification

CAAPS has facilities for changing the conformation of the protein. The two permissible modifications are: *1*) rotating a side chain around the CA-CB bond, and *2*) substituting one side chain for another at a selected residue.

Side chain rotations

Command: ROTAT <picture>, <residue>

ROTAT causes CAAPS to enter its rotation routine for the purpose of rotating the side chain of residue number <residue>, which is currently being displayed in picture number <picture>. Execution of the ROTAT command causes CAAPS to: *1*) display an angle label showing the current position of the side chain relative to the position in which it was originally defined in the coordinate file, and *2*) wait for input from switches or the teletype. The rotation speed and direction are controlled by switches.

CAAPS also will accept input from the teletype during rotations. The permissible inputs are:

Type-in	*Action taken*
"EXIT"	Cease rotation, return to command mode.

"WATCH" Activate the atom contact routines (see SETUP command). Dashed lines will appear between the atoms in contact.

"XWATCH" Deactivate atom contact routines.

<number> Set rotation angle to specified number (degrees).

Side chain substitutions

Command: SUBST <residue number>, <3-letter amino acid name>

SUBST takes a side chain of the type specified by <3-letter amino acid name> and puts it on the <residue number> residue. If the second argument is omitted, the residue returns to its original state with its original side chain.

Atom contact monitoring

CAAPS is equipped to monitor van der Waals contacts between parts of the protein structure undergoing modification. There are two sets of atom contact routines to monitor atom contacts *1*) between a rotating side chain and surrounding parts of the protein, and *2*) between two pictures in relative motion.

CAAPS makes use of special tables in core for atom contact calculations in order to save time. These tables must be initialized by the SETUP command before using the contact routines.

Command: SETUP <residue number>
or SETUP (no arguments)

SETUP initializes the tables in the atom contact routines with the coordinates involved in the calculations. The action taken by the SETUP command depends on which atom contact routine is used.

Rotating-side-chain atom contacts <Residue number> must be the number of the residue that is to be rotated. CAAPS will search the data file for all atoms that could possibly come into contact with a side chain atom of the residue to be rotated. These atoms will be stored in special tables and will be used in computing contacts during rotations. The atom contact routines are activated by typing "WATCH" during the rotations (See ROTAT command), and are deactivated by typing "XWATCH".

Picture-to-picture atom contacts SETUP is typed with no arguments. The user enters two lists of atoms, and atom contacts between the two lists are monitored. The atom contact routines for picture-to-picture contacts are activated by the "WATCH" command and deactivated by "XWATCH".

Command: WATCH

WATCH activates the atom contact routines. Any contacts detected by the routines will appear on the display as dashed lines. Control returns to command mode immediately, but the contacts are constantly monitored until the receipt of an "XWATCH" command.

Other commands

Most of the commands above have complements, e. g., to delete PICTR 1, type XPICT 1, and so on. Pictures can also be "BLINK"ed which is useful for distinguishing parts of a complex display.

Hard copy can be obtained by photographing or filing the screen, or by output to a CALCOMP plotter.

With the stereo lorgnette, pictures can be viewed in true 3-D with rotation. The command is simply DSTER.

It is easy to use. Biochemistry students who were not only ignorant of computers, but suffered from what might be called an aggressive apathy about the machines, began using the system after only a 1-hour introduction to the system. All of them felt that their understanding of large protein structures was enhanced, and felt that computer displays were a great asset in the study of these structures.

I have described the protein display program in some detail, because it does things that are probably recognizable to most biochemists.

Several other projects have made use of the system, in particular our work on nucleic acids and their components. Most of these are of more specialized interest to crystallographers and will only be summarized.

GpC

The structure of the calcium salt of the nucleic acid component guanosine 3′,5′-cytidine monophosphate was solved by energy minimization and display analysis. The molecule is small by nucleic acid and protein standards, but represented a considerable crystallographic challenge.

Fibers

X-ray diffraction studies of the large nucleic acids are usually made on fibers, yielding diffraction photographs. The general parameters (helix pitch, number of residues per turn) can usually be obtained fairly directly, but molecular model building is required. We have designed a program (Bond, Lesk, and Stellman) to construct stereochemically satisfactory nucleic acid models.

The nucleotides are displayed graphically in the correct helical relationship but not joined. The backbone torsion angles are varied to join the nucleotides (via the "bright" bond in the display) while preserving the symmetry relations. The parameter values are displayed during the operation. The final refinement can be done numerically.

Several computer graphics installations have devoted a great deal of attention to the problem of fitting a molecular model into electron density (the computer graphic analog of the Richards Box.) While this is of considerable interest to protein crystallographers, it is a very specialized application and of little direct interest to biochemists and molecular biologists. We are working on this problem to aid in the solution of the structure of the plant lectin wheat germ agglutinin, which is being used as a probe for cell surface changes in tumor development. Preliminary X-ray work is complete and the electron density should be available in a few months.

The aim here is to check, develop and refine these folding proposals, until they can truly be described as algorithms. (Most presently involve nonquantitative qualifiers, such as "usually" or "infrequently".) Although some success has been achieved in alpha helix prediction, and somewhat less in predicting beta sheet, the three-dimensionality of protein structures is crucially dependent on the euphemistically labeled "random coil" regions, and it is in the research that one must go beyond one- and two-dimensional representations and visualize the actual three-dimensional structures predicted, preferably as they are generated. For this, an interactive display is essential.

The rapid increase in the number of laboratories using computer graphics in studies on macromolecules is the best indication of the growing importance of the technique.

Interactive three-dimensional computer graphics is now an essential tool with enormous potential, not only for the visualization of the static structures of large biological macromolecules, but for the still more exciting and still hardly explored dynamics of their interactions with each other. Enzyme activity, allosterism, and protein–nucleic acid interactions are only examples. All biological activity rests fundamentally on molecular interactions, and no static sketches or lists of numbers can compare with direct visualization. Interactive three-dimensional computer graphics is the only way. There is much to be done.

Important contributors to the success of the Princeton University Computer Graphics

Work described include P. J. Bond, S. Broyde, T. Dyott, P. Gund, S. Harbison, B. Hingerty, A. Lesk, J. A. Mervine, S. D. Stellmann, and W. T. Wipke.

REFERENCES

1. Barry, C. D., C. Levinthal, S. A. Ward and M. Zwick. Computer graphics in macromolecular chemistry. In: *Emerging Concepts in Computer Graphics*, edited by D. Secrest and J. Nievergelt. New York: Benjamin, 1968, p. 251–253.
2. Bond, P. J. Computer graphics and macromolecular structures. *Computer Graphics, ACM SIGRAPH* 6: 13–26, 1972.
3. Langridge, R., and E. A. W. Mac Ewan. The refinement of nucleic acid structures. *Proc. IBM Symp. Scientific Computing on Computer-Aided Experimentation* 1965, 133–143.
4. Levinthal, C. Molecular model building by computer. *Sci. Amer.* 214: 42–52, 1966.
5. Newman, W., and R. Sprocell. *Principles of Interactive Computer Graphics.* New York: McGraw-Hill, 1973.

Computer-aided reconstruction from serial sections[1]

C. LEVINTHAL, E. MACAGNO AND C. TOUNTAS
Columbia University, New York, New York 10027

The development of CARTOS, a system for Computer-Aided Reconstruction by Tracing Serial Sections, has been directed by efforts to investigate some rather specific biological questions. We want to study the detailed structure of simple nerve nets so that the anatomy of individual nerve cells and their fibers can be recorded in three dimensions and the morphologically visible synapses connecting them noted. Furthermore, we would like to carry out these morphological studies as a function of time during embryological development in organisms which reproduce clonally so that, except for mutations, all individuals within a clone have the same genes. The specific questions we have been trying to answer are: *a*) To what extent is there variation in the morphology of particular cells when they are examined in different but genetically identical animals raised under identical environmental conditions? *b*) To what extent can morphological variation be induced by changes in the environmental conditions during development? *c*) Within the same animal how precisely does the anatomy of a particular cell match the mirror image of its bilaterally symmetric mate? *d*) Mathematically precise bilateral symmetry must break down when fibers cross the midplane of an animal. Is this breaking of symmetry a feature that occurs in the same way in different organisms? *e*) What is the sequence of development steps leading to the adult structure of the visual nervous system and is this sequence unique and reproducible between animals that have the same genes? *f*) What clues concerning the mechanism by which particular cells make specific connection can be obtained by following the progress of a particular neuron during its development?

The design for a system for computer-aided reconstruction of anatomical structures was started at MIT in 1965. Our initial work on molecular model building (1, 2) had been done at Project MAC using a graphics system that was built at the project and depended on an IBM 7094 to buffer and maintain the display lists. Our second 3-D display system in our laboratory at MIT consisted of a PDP-7 with a DEC display unit as a stand-alone system tied to the time-sharing Project MAC by a 1200 baud phone line. The laboratory was moved to Columbia University shortly after this system was installed.

In the spring of 1967 detailed design and planning was started for a computer graphics facility in the biology department of Columbia University. This led in 1968 to the installation of an Adage Graphics Terminal (AGT-50) and the start of construction of a high-speed bit-

---

[1] Supported by a gift from the RGK Foundation and by Public Health Service Facility Grant #RR00442 (04).

parallel link between our facility and the IBM 360/91 of the Columbia University Computer Center (CUCC). Initially, data transfer between CUCC and the facility was done by carrying magnetic tapes between the two installations. In the spring of 1969 the high-speed link started to function.

The present Adage-360 system is linked through a parallel data adapter on the 360/91. The data transfer is at the rate of about 100,000 30-bit Adage words per second. Three-dimensional objects represented by a few thousand 3-D vectors can be displayed and rotated flicker-free using the Adage hardware matrix multiplier with the display list being maintained in the Adage core. A set of buttons and dials provides user interaction with the Adage.

The design and implementation of the CARTOS operation were based on using the computer as a three-dimensional notebook for obtaining the input of the data to the machine, and then carrying out the analysis in an interactive mode. The first program for recording three-dimensional information was written as a Fortran routine which called various system programs for input and display. Although the first program had some rather awkward limitations it demonstrated the usefulness of the entire system. Individual cells and their fibers could be followed and recorded in the neuropil of a small invertebrate using serial section photographs that had been successively aligned into a film strip on the Image Combiner as described below. The system was functioning in 1970 and the summary description of the design and initial implementation was given in Levinthal and Ware (3).

As with several other applications of computer graphics in biology, the initial implementation showed great promise and created beautiful pictures, but, for various reasons, no questions of substance were answered with reliable quantitative data until considerably later. The basic problem is that interactive graphics must be designed so that an investigator can obtain large amounts of data; however, to be substantively useful it requires that a great deal of time be spent collecting and analyzing the data so that quantitative results are obtained. This means that an interactive system that is to be scientifically valuable must be convenient and simple to use and the investigators must avoid the trap of playing games with the nice pictures.

IMAGE COMBINING

Even with an elaborate computer system for recording data from individual sections the identification and classification of features belonging to different cells in a group of such photographs of nervous tissue can be extremely complicated, especially in regions of profuse branching, as in the neuropil of an invertebrate ganglion. Even with some reasonable landmarks, it is difficult to look at two prints of adjacent serial sections and uniquely determine the corresponding features between them except in trivial cases. This identification is greatly facilitated however if one can superpose the two images and align them so that the corresponding aspects overlap, a procedure long used by astronomers in order to identify specific changes in the maps of regions of the sky. The alignment of such pairs of sky photographs is carried out in general by optically superposing them and alternately illuminating each picture in sequence in a device which has been referred to as a "blinker." This facilitates the alignment on the one hand and enhances small changes on the other by producing a sense of movement on the superposed images of a magnitude that is directly proportional to the discrepancy in the overlap.

The situation for the biologist is complicated by the fact that serial sections usually number in the hundreds or even thousands and any feature of interest may extend over many of these. The logical extension of this technique is to make a motion picture film strip of aligned serial sections in which each frame is a picture of one section. These film strips can be studied for any feature of interest by projecting the film strip sequentially in a movie projector. Such films were first made some 70 years ago in Germany, but due to various technical difficulties the technique has not been used extensively until recently. When such film strips are well made one gets the illusion of three dimensions and the effect that one is "walking through the structure" as the film is projected at near movie speeds.

We have designed and constructed in our laboratory various machines for making film strips from serial sections, incor-

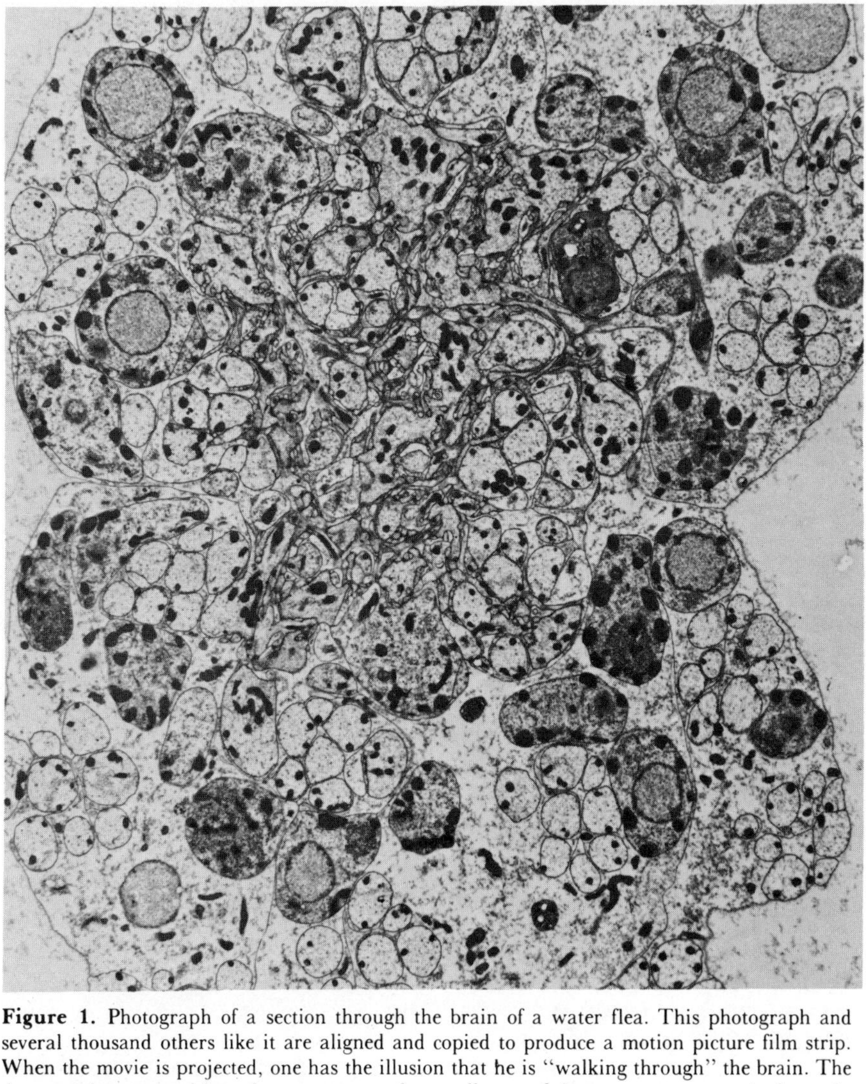

Figure 1. Photograph of a section through the brain of a water flea. This photograph and several thousand others like it are aligned and copied to produce a motion picture film strip. When the movie is projected, one has the illusion that he is "walking through" the brain. The data are then entered into the computer and a small part of the computer output is shown in Figs. 2 and 3.

porating details and ideas from earlier and simultaneous work of other laboratories. Since our work involves material that must be studied on both the optical as well as the electron microscope we felt it necessary to design a device, which we called an "image combiner," that could photograph onto film strips the negatives of sections taken in either microscope. An earlier design which duplicated one by Zimmerman (7) consisted of two microscopes connected by a comparison tube which allowed one to see simultaneously the images from each microscope. By placing even-numbered sections under one microscope and odd-numbered sections under the second, one could align adjacent sections with each other by moving the corresponding microscope stage with respect to the previous one. After each alignment the sections were photographed and the process repeated. Besides the fact that films could only be made in this manner from light microscope sections there were some inherent problems with the design that, in our hands, limited its resolution.

The next design was specifically made for filming 70 mm electron micrographs. This device consisted of mirrors, some half silvered, that were used to mix the images of two subsequent photographs, two mechanical stages that held the negatives in position, and appropriate viewing, photographing and lighting equipment. A number of useful films were made on this device; however, it had a limiting feature which reduced the resolution that could be obtained on the final film strip. The filming was done through the same mirrors as the viewing for alignment and, because the size of the negative necessitated a rather large instrument, the distance between the copying lens and the negative was excessively large for obtaining high resolutions. New arrangements were then considered that would permit us to bring the copying lens closer to the negative even at a loss in viewing area, so that the finer details in the electron micrographs could be studied with the 35 mm filmstrip. Various designs were studied, and the one that we have been using for some time arose from a suggestion by Dr. Sidney Brenner (personal communication), in which a rotating turntable is set up with three positions; two of these are used for aligning the negatives and a third is used for the photography. Our present machine, which has been used for essentially all the substantive work that has been done in our laboratory, involves a mechanical system for maintaining the appropriate orientation of the individual negatives as the turntable is rotated 120° at a time. This instrument is mechanically a bit elaborate and it requires rather high precision alignment since if this is not done the resultant film strips will show a cumulative drift in one direction. Although this drift can in principle be corrected later in the computer it is sufficiently awkward so that we are currently designing a new version that should be mechanically simpler. Another version of the "image combiner" which is mechanically very simple although not inexpensive involved viewing the sequential images and carrying out the alignment with the use of a video camera, monitor and video disc. Briefly, a negative was imaged and recorded on the disc, then a second negative was placed on a single stage and imaged in the video camera. The camera image and the recorded one were then alternately displayed on the monitor and the position of the currently viewed negative was moved until alignment was obtained. A working prototype of this combiner was made in our laboratory; however, we decided not to use it for further work at this time because the resolution of our video system was not sufficiently high for routine work with the electron micrographs.

At present the number and variety of devices being used in our laboratory and elsewhere for image combining is large. However, the basic principle is the same as that used in the earlier ones. The usefulness of this process, even without

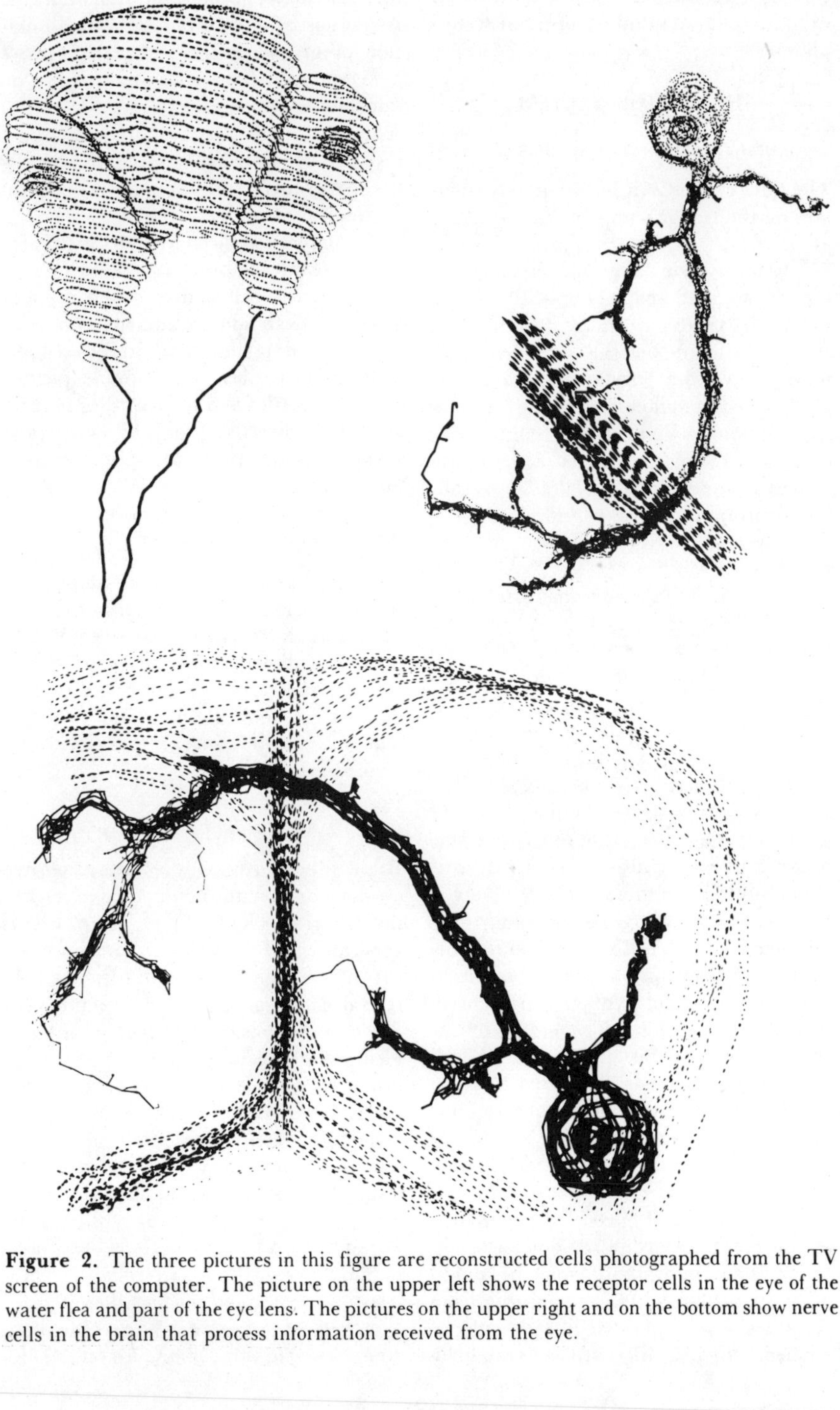

Figure 2. The three pictures in this figure are reconstructed cells photographed from the TV screen of the computer. The picture on the upper left shows the receptor cells in the eye of the water flea and part of the eye lens. The pictures on the upper right and on the bottom show nerve cells in the brain that process information received from the eye.

computer recording, is clear when one is examining large numbers of serial section photographs.

THE CARTOS SYSTEM

Computer representation of data

The bulk of the CARTOS software operates on the Adage computer in the stand-alone mode or on the coupled 360/91–Adage system. The output image on the Adage oscilloscope is optically mixed with the projected image of the film strip. A cursor on the oscilloscope is moved by the observer using the "mouse," a handheld device for indicating position in X and Y. In addition, the observer controls the single frame movie projector (and therefore the Z coordinate) through the Adage. Thus, the observer can effectively "draw" in three dimensions.

The CARTOS programs use these three-dimensional drawings to generate and operate on four types of 3-D data files: *a*) A *brain file* contains over-all information about the movie of a brain (magnification, section thickness, etc.). Normally, there is one brain file for each movie. *b*) A *nerve file* contains all information that is recorded about a particular nerve cell. Each nerve file consists of four areas: a fixed-size header, a contour area containing the contours of the cell body, a tree area containing the topology and geometry of the cell's branching structures, and a synapse area in which locations and types of synapses are stored. Original nerve files are generated by the interactive tracing programs (REC, GTREC), but other programs both on the Adage and on the 360 are also capable of generating such files by editing or otherwise processing traced information. *c*) A *nucleus file* contains the locations and sizes of all cell nuclei (approximated as spheres) in a certain region of a brain. In addition, information may be stored associating pairs of nuclei resulting from cell division, so that cell lineage may be studied. Nucleus files are generated by the interactive program NCL (on the Adage) or by editing and data manipulation programs on the 360. *d*) A *picture file* is simply a display list and contains no information about the picture it represents. Picture files are usually temporary; they are generated by editing or processing programs, and they never contain original, traced data.

A completely traced film strip is represented in the computer by one brain file, one or more nucleus files, and a number of nerve files. Editing and manipulation of this raw data may result in the creation of more such files, or of simple picture files. Nerve files and nucleus files may be displayed directly by CARTOS programs, without prior conversion to picture files.

Adage stand-alone programs

The CARTOS Adage stand-alone software consists of interactive data recording programs, interactive display (playback) programs allowing more advanced graphical manipulation, and a variety of general utility and file-handling routines. The most important programs are described briefly below:

Recording programs

Brain files and nerve files are generated on the Adage with the interactive recording program REC. The brain file is generated first by typing in some numerical information, and nerve files are then recorded one at a time. The name of a nerve file consists of two characters identifying the brain it belongs to, followed by up to three digits representing the unique "nerve number" of the corresponding cell. Thus, the Adage system's file directory can be used to establish a simple data base for nerve file retrieval which is easily usable by all other Adage programs. (Similar naming conventions simplify the retrieval of brain and nucleus files.)

A nerve file need not be recorded in a single session; it may be saved on the disk in "partial" status and its tracing may be resumed at a later time. When completed, its status is changed so that it may not be later inadvertently modified. REC keeps track of all nerve files traced for a certain brain, informing the tracer about the status of the file he selected to trace ("partial," "complete," or "new file").

While tracing the nerve file, the user may enter in any sequence, and any number of times, each of the three recording modes: "contour," "tree," and "synapse." These modes are selected through function switches, and each mode assigns different functions to the three "mouse" push-buttons. Thus, in contour mode the button functions are (invisible vector/visible vector/delete last vector), in tree mode they are (node/continuation point/terminus), and in synapse mode (pre-synapse/post-synapse/ambiguous).

In contour mode, the outline of the cell is traced for each movie frame as a series

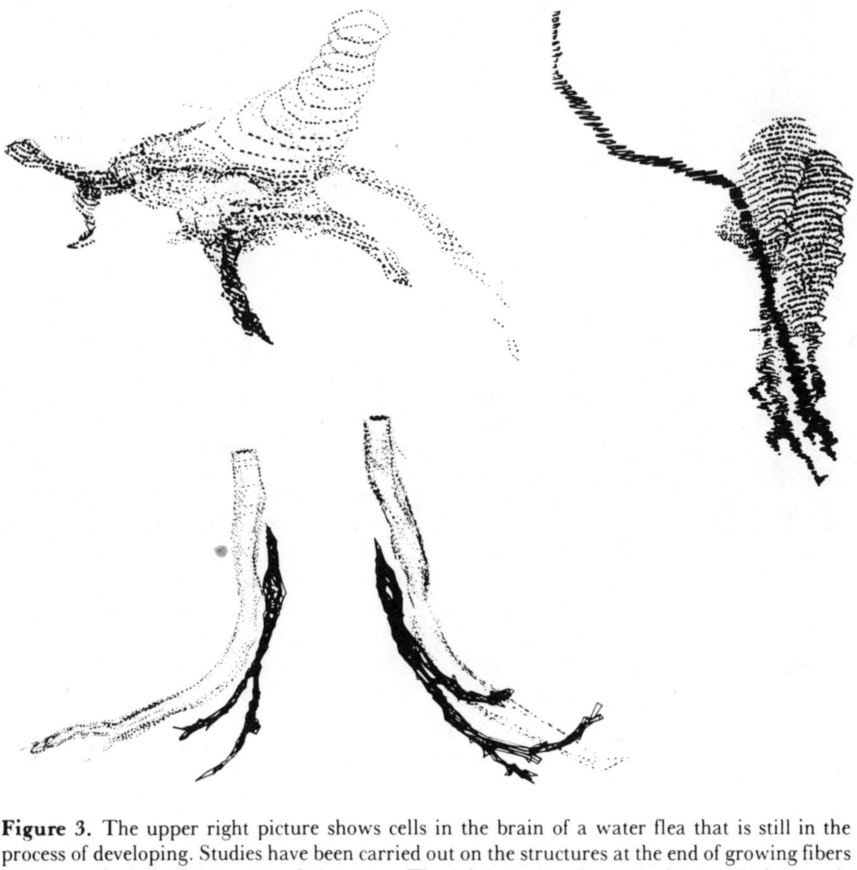

Figure 3. The upper right picture shows cells in the brain of a water flea that is still in the process of developing. Studies have been carried out on the structures at the end of growing fibers which are finding their way to their target. The other upper picture is the terminal synaptic region of a receptor cell in an adult Daphnia. The lower picture shows four cells in the brain of a small fish. This fish, a tropical Molly, is used because animals can be obtained easily that are genetically identical. Variations are being studied in the brain structure of what are effectively "identical twins."

of line segments. The tracking cursor may be either a dot or a circle whose radius is controlled by a dial. Special functions allow the automatic generation of a sphere defined by the center and radius of the tracking circle (a feature which expedites the tracing of approximately spherical shapes, e.g., the cell nucleus) and the automatic duplication of a previously traced contour in a new frame.

While tracing, it is possible to display only that part of the already traced information that lies in the current frame or in a three-dimensional slice of variable thickness that is centered at the current frame. This "windowing" feature makes it possible to avoid visual clutter while keeping in view, if it is desirable, the closest already traced features.

Synapses are recorded as visible points with a one-letter label identifying their type. The tracing of tree structures is facilitated by the fact that the program always remembers the most recently recorded "unsatisfied" node, and points it out to the tracer as soon as a terminus is entered. The projector is then automatically returned to the frame in which that node was traced, and an invisible vector to the node is entered. The most recently traced branch of the tree always appears brighter, and the projector may always be moved to the last traced contour or tree point, so that the resumption of tracing after a long interruption presents no problems of forgetting where one had stopped.

In addition to the three recording modes, there is a "view" mode which allows the tracer to rotate (in 3-D) or blow up the traced picture. It is also possible to oscillate the entire picture about a vertical axis going through the current position of the cursor.

The current file is normally updated several times during tracing by pressing a push button. An automatic update is also performed whenever the program performs a reallocation of free core (it should be noted that a rather elaborate core allocation scheme is required to allow the simultaneous tracing of contours, trees, and synapses, and the maintenance of a variable size push-down stack, while each of these areas remains contiguous in core and no memory is ever wasted).

Display and graphic manipulation programs

Displayable files can be grouped and their display controlled separately by push buttons. Each group may be manipulated independently while the entire display is being rotated or translated with respect to laboratory coordinates. These functions make it possible to isolate or point out features in a collection of traced cells, or to superimpose and compare corresponding cells from different animals. Contours, trees, and synapses may be turned on or off either within each group or for all groups with the use of push buttons. It has been found that the extensive push button controls of GSHOW provide a very powerful interactive tool in morphological investigations by making possible the detection of patterns of cell interaction in complicated three-dimensional configurations.

Another display program makes it possible to study the extent of bilateral symmetry present in a traced region, and to determine the exact position of the midplane. This is accomplished by moving the displayed picture with respect to a vertical plane and superimposing the half of the picture lying on one side of this plane with the mirror image of the other half. Flashing each of these images on and off alternately generates an illusion of motion which (as is the case with the image combiner) is easily minimized using interactive controls.

Programs using Adage-360 coupled system

At present, all data analysis requiring significant amounts of core or computation is done on the 360/91 system. The high-speed core-to-core link connects the

360 with our Adage, making it possible to run interactive Adage programs that communicate with 360 programs running simultaneously. The communications software for the two machines has been implemented in three levels. The *first* level involves simple transfer of binary and text data from core to core. The *second* level includes a coded message scheme which allows 360 programs to treat the Adage system as three independent peripheral devices: an interactive user console, a display unit, and a disk. At this level, the user may type commands directly to the 360, which then may read or write Adage disk files and show pictures on the Adage screen. The *third* level communications software allows direct interaction of 360 programs with the tracing program REC or the display program GSHOW on the Adage.

An example of the use of third level software with REC is the 360 contour-fitting program. The tracer proceeds with REC as in the stand-alone mode, but has the option of tracing fewer contours (for example, one contour every tenth frame) and letting the 360 program compute and store the intermediate contours automatically; the tracer may inspect the computed contours by viewing each one on top of the corresponding movie frame, and may then accept or reject them—in the latter case, he traces some of the intermediate contours himself and then asks for a new fit.

CONCLUSION

During the past 2 years the system has been used extensively for 3-D reconstruction of nerve nets in the brains of three organisms which reproduce parthenogenetically so that clones of isogenic organisms are obtained. Three papers have been published on the work with the water flea, *Daphnia magna* (4–6), and manuscripts are in preparation on work with the rotifer *Asplachna brightwelli* (I. Hammerman, unpublished) and the fish *Poecilia formosa* (F. Levinthal and E. Meyerowitz, unpublished). We will not attempt to summarize here the substantive results of this work, but it is important to note that several of the conclusions that have been reached could not have been deduced, we think, by simply looking at prints of the serial section photographs. The CARTOS system has already been of major usefulness in that it has allowed a number of observations to be made in partial answer to some of the questions enumerated above. It now seems clear that the technique is a necessary adjunct to many types of morphological analysis.

We have recently designed and implemented a new tracing terminal which is based on a small PDP11/05 in a DEC GT-40 display unit. This should make the CARTOS system available to a much larger group of investigators; the total hardware, including the computer display, projector, digitizer and interfacing costs about half the price of an electron microscope. It therefore seems likely that this type of computer-aided reconstruction from serial sections can become a routine part of the work in many electron microscope laboratories.

Many people have participated at various stages of the development and implementation of the computer system here discussed. Particularly important were the following: Steven Ward; Jack Hahn; Lou Katz; Reidar Bornholdt; Randle Ware; William Ross; and Paul Green.

REFERENCES

1. Landridge, R., and E. A. W. McEwan. *IBM Symp. Computer-Aided Experimentation.* 1965, 133–143.
2. Levinthal, C. *Sci. Am.* 214: 42–52, 1966.
3. Levinthal, C., and R. Ware. *Nature* 236: 207–210, 1972.
4. LoPresti, V., E. R. Macagno and C. Levinthal. *Proc. Natl. Acad. Sci. U. S.* 70: 433–437, 1973.
5. LoPresti, V., E. R. Macagno and C. Levin-

THAL. *Proc. Natl. Acad. Sci. U. S.* 71: 1098–1102, 1974.
6. MACAGNO, E. R., V. LOPRESTI AND C. LEVINTHAL. *Proc. Natl. Acad. Sci. U. S.* 70: 57–61, 1973.
7. ZIMMERMAN, M. H. *Science* 152: 71–72, 1966.

Computer-based modeling and interpretation in medicine and psychology; the Rutgers Research Resource

SAUL AMAREL
Rutgers University
New Brunswick, New Jersey 08903

MODELING APPROACH

The Rutgers Research Resource on Computers in Biomedicine[1] was established about 3 years ago with the objective of bringing advances in computer science —particularly in artificial intelligence and in interactive data base systems—into problems of biomedical inquiry. The main thrust of the Resource is to develop ways of using the computer as an integral part of the inquiry process. Our thesis is that the computer can be made to play an important role in this process if its potential for symbol manipulation is used to provide 'intelligent assistance' in the organization of scientific knowledge and in its utilization for problem solving.

An essential part of the Resource effort is oriented to methodological problems of modeling and to the development of computer-based systems for the construction and use of models in a variety of tasks.

The use of computer models can be of considerable help in understanding a system whose basic structure and relationships are still not completely known. A researcher is likely to gain insight into the poorly understood parts of such a system by specifying a model in which certain theoretical ideas are incorporated, and by studying it on a computer. The computer can become a powerful research tool if it is used not only for testing given models but also for *creating* models, for updating them, and for offering advice as to how to restructure them in the light of available evidence.

We encounter in this area several basic methodological problems: how to choose a modeling approach; how to establish relationships between different models of the same domain of phenomena, and how to maintain consistency in these relationships when some of the models change; how to choose representations for models; how to form, validate and update models. We also have important implementation problems, involving the choice of computer systems and languages for handling models. Flexibility in creating models, changing them, and using them in different processes is an important requirement for the use of models in research. This requirement strongly influences the mode of representing models in computers and the systems used to manage them. Our general approach is to separate the representation of a model from the programs that use the model—for simulation, interpretation, instruction, and so on. The problem of choosing a representation for a model is significant both from the theoretical point of view (scope, expres-

[1]The Resource is supported by N.I.H. Grant #RR-643.

sive power) and also from the pragmatic point of view (ease of changing models, and efficiency of their utilization in different computer processes).

These problems are important subjects of study in Artificial Intelligence (AI) at present. Artificial Intelligence is broadly concerned with problem solving processes and with their implementation in computers. Questions of knowledge acquisition, representation, and utilization in problem solving processes are central to current work in AI. These are difficult questions that are closely related to the problems of modeling mentioned above. Some of them have been studied in the context of controlled problem solving situations (1, 2, 3). We believe that we will gain much by studying these questions in the realistic setting of scientific work in biomedicine. Of particular interest to us are situations where knowledge of different types (forms, completeness, validity) is available for decision making —such as in medical diagnosis and treatment. We have chosen therefore, to concentrate on problems of model-guided decision making in specific areas of medicine, and also on problems of scientific inquiry in social psychology and in ecosystems that involve various types of model building tasks.

In addition to its significance in theory construction, a computer-based framework that can provide effective assistance in the creation and utilization of models will play an important role in reducing the gap between science and practice in areas such as medicine and psychology. Medicine is primarily concerned with diagnostic decisions and with the design of courses of action for the treatment of patients. Knowledge of different types —some of which comes from research in the life sciences—enters into the reasoning processes of the medical decision maker. By studying medical problem solving in a computer, in the context of models that embody physiological/pathological knowledge of body processes, we expect to obtain a better understanding of how this knowledge can be used in the reasoning processes that lead to medical decisions. If we establish systematic ways of using such models in medical problem solving then new knowledge about a body process, a basic mechanism of a disease, and the like can be immediately related to clinical observations and actions, and thus it can be used and tested in medical practice. The contribution of the computer system in these situations is in handling the complexity of the relationship between a large and changing body of relevant general knowledge and decisions in special cases. We have a similar situation in some areas of psychology, where the computer is starting to provide a realistic means for relating complex bodies of proposed theory to experimental findings.

The computer systems that we are developing and studying in the Resource are characterized by the following capabilities: *i*) convenient man–machine interface, *ii*) flexible facilities for representing and managing knowledge in a specific domain, and *iii*) procedures for creating and modifying models, and programs for using models in problem solving. These systems are being developed jointly by computer scientists and life scientists who collaborate in both the methodological and implementation aspects of system design. A key element in these collaborations is the choice of 'discipline oriented' projects that provide a concrete basis for working interactions between people in computer science and people-with-problems in biomedical disciplines. Within the Resource, the value of a project as a vehicle for interdisciplinary coupling and stimulation is as important as its intrinsic scientific merit. Another important aspect of a 'discipline oriented' project is its value as a paradigm for the computer-aided approach to inquiry that we are developing.

OUTLINE OF ACTIVITIES IN THE RUTGERS RESEARCH RESOURCE

At present, work in the Resource is organized in six active areas of study; in each area there are one or more projects. Work in three of the areas is in 'discipline-oriented' projects. These areas are: Medical Modeling and Decision Making, Modeling of Ecological Systems, and Models of Belief Systems. The three remaining areas are devoted to 'core' work in computer science which is closely coupled to the other 'discipline-oriented' studies. These areas are: Representations and Modeling in Artificial Intelligence, Descriptive Systems and Data Structures, and Languages and Interactive Systems.

The Resource community includes about 35 researchers and it consists of an active group of members, and of two related groups of associates and collaborators. The main work of the Resource is carried out by its members. In general, associates are less actively involved, but they play an important role in communication activities, and in the development of new collaborations. Most of the members and associates come from different academic units of Rutgers (Computer Science, Psychology, Zoology, Ecology, EE-Biomedical Engineering); some are associated with other institutions in the New York–New Jersey metropolitan area (Mount Sinai School of Medicine in New York, East Orange VA Hospital, New Jersey College of Medicine and Dentistry—Rutgers Medical School). Resource collaborators are usually affiliated with institutions in other parts of the country. They form national collaborative networks of investigators, and they interact with members of the Resource on specific projects. A major mode of interaction between collaborators and members is via common access (through communication networks) to the computer facilities of the Resource, and joint work on program development and testing. The nucleus of a network of collaborators in the study of glaucoma is now being established. It includes investigators from Mount Sinai School of Medicine in New York, Washington University in St. Louis, and Johns Hopkins University in Baltimore.

The idea of collaboration at a distance in computer-based projects is becoming increasingly feasible with recent advances in time sharing and computer networking. National computer networks such as the Arpanet and Tymnet provide the technical basis for realistic resource sharing. Our plans are to use a PDP-10 at Rutgers, as well as the PDP-10 of the SUMEX-AIM national resource at Stanford, in conjunction with appropriate network linkups, as the basis for national collaborations in the glaucoma project and in other discipline-oriented projects. Such collaborative activities will bring together Resource investigators who are concentrating on methods of computer-based inquiry in given fields with selected scientists—users in the country who are key people in these fields. This mode of work promises to provide effective mechanisms for perfecting, utilizing, and disseminating the developments in the Resource.

It is conceivable that network-based collaborations of the type developed in our Resource will have a strong effect on the pattern and productivity of research in the life sciences. Computer networks coupled with artificial intelligence methods promise to provide more than powerful research tools; they have the potential of creating new frameworks for research and of introducing new transfer mechanisms between research and practice.

STATUS OF RESEARCH IN THE RUTGERS RESEARCH RESOURCE

In the area of *Medical Modeling and Decision Making*, our research has concentrated on a modeling approach which is based on causal networks and is well suited for prediction, exploration and clinical decision making (diagnosis and treatment). This approach has been applied to the implementation of a flexible and comprehensive consultation program for glaucoma. The glaucoma program is being further developed at present with the emphasis on strategies of therapy. Work on the consultation program has been received enthusiastically by eminent investigators in the field of glaucoma, and this led to the development of a national network of collaboration in glaucoma. Testing, improvement and refining of the glaucoma consultation program are now being carried out in the framework of this national network of collaborators. One of the outstanding problems in this area is how to increase the assistance that a computer can provide in the process of updating/improving the causal model of glaucoma on the basis of clinical experience as well as by assimilating new scientific knowledge on the physiology and function of the eye. Other projects in the medical area involve modeling in hematology as well as in other fields of ophthalmology (refraction, strabismus).

Our research in the area of *Modeling Ecological Systems* has concentrated on the study of the kinetics of nutrients in microecosystems and on succession phenomena in the phytoplankton community. The emphasis is on mathematical models in the form of ordinary or partial differential equations. Computer facilities to assist in the model building process have been implemented. One of the interesting problems in this area is how to design a computer system for (semi) automatic formation of compartmental models. This problem requires an approach that combines methods of artificial intelligence and numerical computing.

Work in the area of *Belief Systems* has led to the development of a computer-based theoretical framework for the common sense interpretation of social episodes. A large part of this framework has been implemented. It integrates recent work in artificial intelligence on reasoning about actions with empirical work in psychology on the perception and understanding of social behavior. The model is being continously extended, and it is being used in studies of developmental aspects of social interpretation as well as in the study of special communication paradigms, such as the psychiatric interview. An important problem in this area is to devise empirical procedures for testing the computer models.

Our work on *Representations and Modeling in Artificial Intelligence* involves applications and extensions of AI methods to the 'discipline oriented' projects in the Resource, and also identification and study of AI problems that are significant to biomedical research and practice. Our studies include schemes for diagnostic problem solving and for reasoning about actions; question answering systems; theory formation processes; problems of shifts between representations in problem solving processes; problems of representing and processing event sequences; and problems in the area of natural language understanding. A powerful program for natural language processing (CHRONOS) has been developed, and it is being adapted to various systems that are being developed in the Resource with the purpose of providing them convenient man–machine interfaces. A significant characteristic of CHRONOS is its ability to specialize to different domains of discourse.

In the area of *Descriptive Systems and Data Structures* our work has concentrated on the development and implementation of a metadescription system (MDS) which can be used to specify and implement specific knowledge-based systems for different domains of discourse. We expect MDS to provide a powerful system-building tool for computer-based scientific work in different areas. At present we are contemplating the application of MDS to the implementation of model-based systems in the medical area.

Our work in the area of *Languages and Interactive Systems* has concentrated on systems for establishing and processing language definitions and also on the development of flexible parsers for these languages. Research on these parsers has extended to more general processes of interpreting event sequences in terms of given systems of rules and also to problems of grammatical inference that have similarities with certain problems of diagnosis and theory formation that we are studying in the Resource. A flexible parser that was developed for syntactic analysis has been adapted for the implementation of interpretation processes in the area of Belief Systems.

WORK ON COMPUTER-BASED INTERPRETATION PROCESSES

One of the key processes in scientific inquiry is the *interpretation* of empirical data in the light of a given body of theoretical knowledge. Much of our work in the Resource is concerned with these processes and their implementation in computers. In the following we shall outline our approach to computer-based interpretation processes in the areas of glaucoma and social causation. In each case we will present sufficient background to permit discussion of the main ideas involved. Readers interested in a fuller discussion of the research in glaucoma should consult the references (4, 10–14, 17); for a fuller discussion of the research in social causation, they should consult the references (15, 16).

Modeling and diagnosis in glaucoma

Glaucoma is a disease of the eye characterized by increased intraocular pressure leading to degeneration of the retinal nerve net, excavation of the optic disk, and atrophy of the optic nerve. These factors are manifested by progressive and irreversible loss of vision culminating in blindness (9).

Several reasons underlie our choice of this disease as a prototype for the development of modeling methods and model-based consultation programs for diagnosis and treatment. Knowledge about glaucoma can be described adequately in terms of processes in a limited body subsystem (the eye). There are sufficiently different types of glaucoma, with different indicated therapies; the ability to make a diagnostic distinction between the types is clinically important. The mechanisms of glaucoma relevant to these distinctions (various abnormalities of flow of the aqueous humor) can be represented to a first approximation as sets of causal statements. This approximation is useful clinically, and there is evidence (from text descriptions and discussions with ophthalmologists) that clinicians employ a strong component of causal justification in their diagnostic reasoning. Early detection and treatment of the disease (which is a leading cause of blindness in this country) is essential to preserving a patient's sight. Such detection involves a subtle process of inference for which a detailed model of possible courses of the disease is extremely valuable.

A causal network of glaucoma has been built by Dr. Kulikowski and his collaborators in our Resource (11, 12, 17) in close consultation with Dr. Safir—an ophthalmologist from the Mount Sinai School of Medicine who has been an active member of the Resource for over two years. The causal statements in the model establish relationships between 'deviation states' of the eye. A state is a specific feature of the anatomy, physiology or function of the eye—such as 'very highly decreased anterior chamber size.' Deviation states are states chosen to describe the processes of glaucoma in the form of causal relationships. The notion of causation is much stronger than that of association or probabilistic coupling. It establishes a time ordering between deviation states, since effects cannot precede their causes. In glaucoma, as in other diseases, it is very difficult to establish a strict statement of causation between two deviation states. This is due to lack of complete knowledge about the processes involved—which may result in inappropriate choices of deviation states (sometimes due to a poor choice of resolution in the description of a process) and/or in uncertainties about relationships between states. Thus, each causal statement in the glaucoma model has a number associated with it (the 'strength of causation') which expresses the degree of certainty in the causal bond.

A causal statement in the glaucoma model has the general form,

$$A \xrightarrow{\alpha} B$$

where A, B are deviation states or logical combinations of deviation states, and α denotes the strength of the statement 'A causes B.' This is a simplified version of statements in the model—in which additional constraints and time relationships are specified. For a more complete description of the model see (11, 17). Six levels of strength of causation are used in the model, and they are given numerical values between 0 and 1, with the lowest value interpreted as 'rarely' and the highest interpreted as 'always.' An example of a causal statement, expressed in English, is the following:

> A very highly decreased anterior chamber size causes almost always both a decreased angle between the iris and the trabecular meshwork and a decreased open area between the iris and the lens.

Causal statements are similar to production rules of the type used in the specification of probabilistic grammars.

The causal model for glaucoma consists of a set of 'connected' causal statements that are structured in the form of a directed graph without loops—where time (and the evolution of disease) flows in the direction of the arrows. A simplified causal network structure for the description of glaucomas is shown in Fig. 1. Only the principal causal chains are shown, and the nodes stand for simplified deviation states with symbols ↑ and ↓ used to denote high and low magnitudes respectively.

A path through the causal network represents a specific disease process, and it defines a class of glaucomas with a specific primary cause (etiology). A class of glaucomas may also be characterized by a combination of connected paths in the causal network, representing a combination of processes that emanate from several primary causes. Pathways that are incomplete, i.e., they do not end in a terminal node in the causal network, represent intermediate stages of glaucoma.

A diagnostic decision for a given patient consists in determining that the patient has a glaucoma of a certain class, and that the glaucoma has evolved up to a certain stage. This is equivalent to associating with a patient a pathway (or a combination of pathways) in the causal network. Such a decision is to be based on tests performed on the patient and on the knowledge embodied in the causal network.

The causal model of glaucoma that is currently implemented on the computer is mainly based on information from two sources: the text by Becker-Shaffer (9) and consultation with Dr. Safir. This model will continue to change on the basis of *a*) new general information (in the form of new causal statements or changes in existing statements) obtained from clinicians, *b*) information about the performance of the current model in clinical environments, and *c*) utilization of scientific information about underlying physiological processes occurring in glaucoma—e.g., processes controlling aqueous flow. Utilization of new information about underlying processes is likely to result in redefinitions of states and in modifications of causal structures in the model.

The best validation of the causal model

would be by testing it in a large number of clinical cases. A successful test of the model in a specific case amounts to a successful utilization of the model in making medical decisions, e.g., diagnostic decisions, in the given case.

Thus, the *process* of diagnostic decision making plays an important part in two significant situations: *i*) in the direct *application* of the model to clinical decision making, and *ii*) in processes of *validating* and restructuring the model—more generally in helping glaucoma researchers to organize and improve the available knowledge in their field.

The diagnostic process that was developed in the context of the causal model (12–14, 17) can be conceived as a process of interpretation within a given system of rules, where the reasoning proceeds in a general 'bottom up' direction, from nodes in the causal network that correspond to observed states in a patient to other nodes higher in the network that represent possible causes of the observed states. The problem is to *complete* a pathway (or a combination of pathways) in the causal network that covers all the nodes that represent observed states in a patient. Starting from a small set of nodes in the network that correspond to confirmed states in a given patient (obtained from a set of initial tests), additional tests are selected sequentially as part of an overall strategy to infer—with high certainty and with minimal cost—the network pathways that represent a diagnosis for the patient.

The strategy used in this process of interpretation can be described as a *probabilistic parser* which proceeds as follows: All parses that are consistent with a given set of confirmed nodes (they correspond to states that are established, with a high degree of confidence, by tests) are generated; parses are ranked by a likelihood measure that depends on confidence measures of tests and on strengths of causal links on the model; and a small number of parses (one or two) of high likelihood measure are selected as proposed diagnoses. This type of exhaustive approach is possible in our case because of the existence of a causal model of disease that covers explicitly, and within a limited amount of network complexity (an order of about 100 state nodes), the space of all relevant disease descriptions.

In situations where the complexity of a causal network is much larger than the complexity of our present glaucoma model, it would be reasonable to consider a 'top–bottom' strategy of interpretation, where hypotheses about causes are selectively generated, and their consequences are followed and evaluated against the specific information obtained from tests. This approach is similar to the reasoning used in the DENDRAL system at Stanford, where the task is to interpret mass spectra on basis of knowledge about chemistry and mass spectroscopy (5, 7).

A major problem in the design of these interpretation schemes is how to effectively use the available information to obtain selectivity of hypothesis generation. This is a central problem in AI; its study in the context of medical diagnosis promises to bring new insights into possible ways of handling it.

For a given causal model, suppose that we have a new body of clinical information (in the form of recorded traces of disease processes) that is inconsistent with the model. The parsing process that is used to obtain a diagnosis in terms of the given causal model can provide the basis for a system that will propose changes in the model—to satisfy the new clinical information.

The problem of model modification is conceptually related to the following problem. Given a grammar in the form of a set of production rules, and given a set of strings that are 'almost derivable' from the grammar (i.e., they are not derivable from the given grammar but they can be derived from a grammar that deviates 'slightly' from the given grammar) find the deviation grammar from which the set

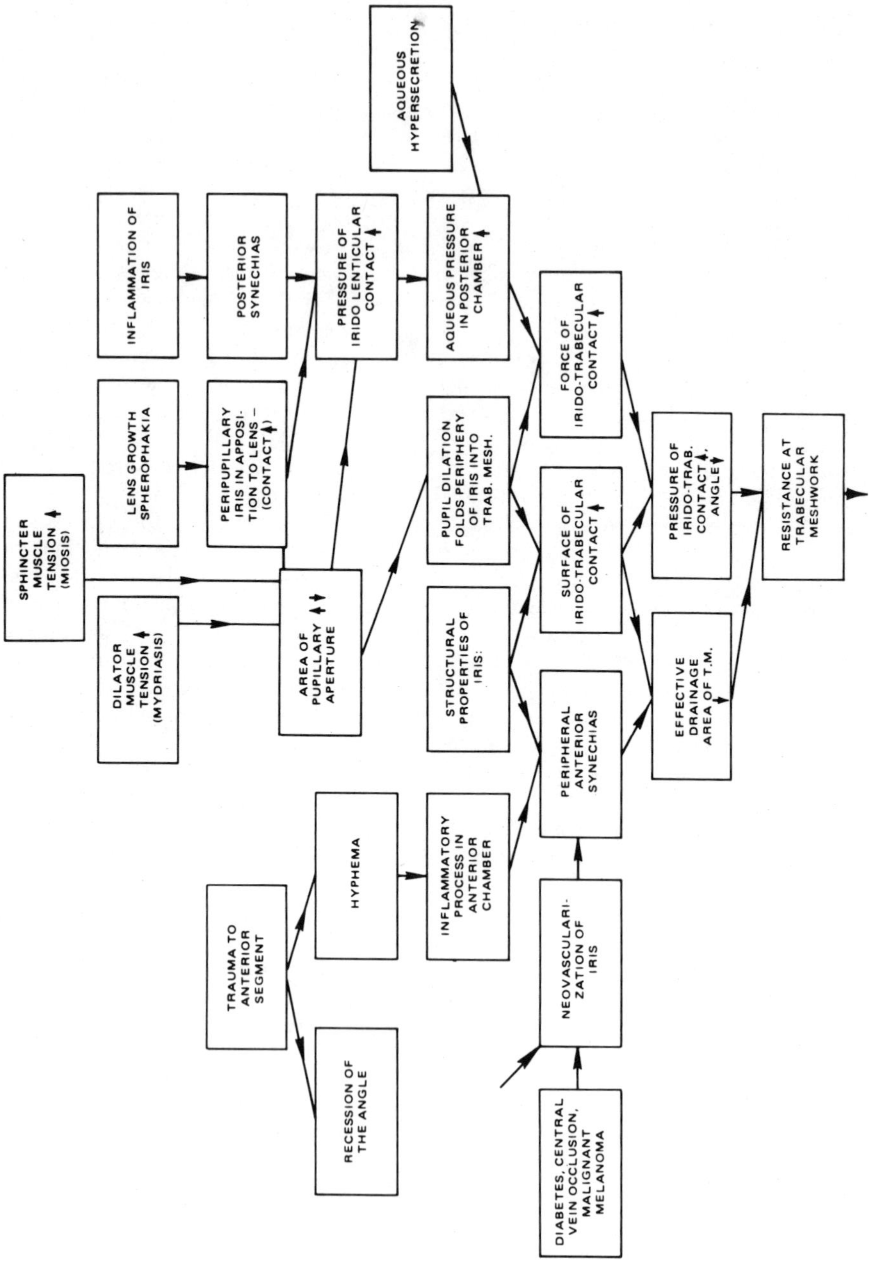
AQUEOUS HYPERSECRETION
INFLAMMATION OF IRIS
POSTERIOR SYNECHIAS
PRESSURE OF IRIDO LENTICULAR CONTACT ↑
AQUEOUS PRESSURE IN POSTERIOR CHAMBER ↑
LENS GROWTH SPHEROPHAKIA
PERIPUPILLARY IRIS IN APPOSITION TO LENS – (CONTACT ↑)
PUPIL DILATION FOLDS PERIPHERY OF IRIS INTO TRAB. MESH.
FORCE OF IRIDO-TRABECULAR CONTACT ↑
SPHINCTER MUSCLE TENSION ↑ (MIOSIS)
SURFACE OF IRIDO-TRABECULAR CONTACT ↑
PRESSURE OF IRIDO-TRAB. CONTACT ↑, ANGLE ↓
RESISTANCE AT TRABECULAR MESHWORK
DILATOR MUSCLE TENSION ↑ (MYDRIASIS)
AREA OF PUPILLARY APERTURE ↓↑
STRUCTURAL PROPERTIES OF IRIS:
EFFECTIVE DRAINAGE AREA OF T.M. ↓
PERIPHERAL ANTERIOR SYNECHIAS
HYPHEMA
INFLAMMATORY PROCESS IN ANTERIOR CHAMBER
TRAUMA TO ANTERIOR SEGMENT
NEOVASCULARIZATION OF IRIS
RECESSION OF THE ANGLE
DIABETES, CENTRAL VEIN OCCLUSION, MALIGNANT MELANOMA

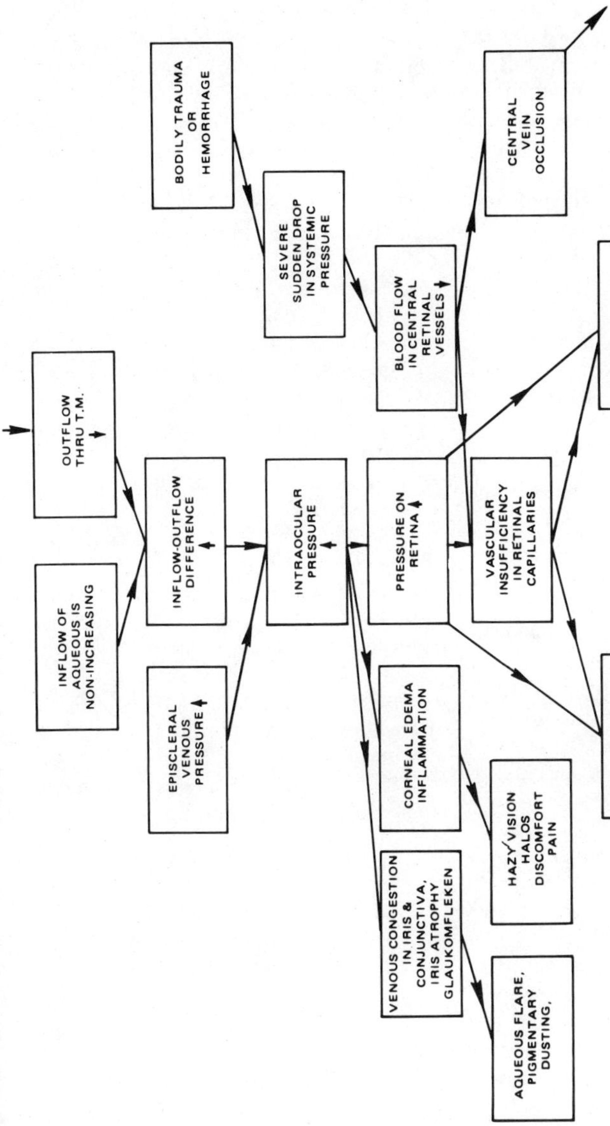

Figure 1. Schematic causal network for glaucoma.

of strings is derivable. This is a grammatical inference problem of the type studied by Dr. Fabens in our Resource. Fabens has developed a flexible relaxation parser that can be used to infer deviation grammars (6).

Some of the ideas developed in the grammatical inference problem will be useful in the problem of causal model restructuring. However, the special knowledge that exists in the medical domain (i.e., a causal link has a meaning in terms of underlying physiological processes) and its special form (i.e., the blending of deterministic and nondeterministic information) may require a completely different approach to the process of causal model verification and restructuring. This is an interesting area of work that we have just started to consider.

Modeling and interpretation in the development of a theory of intention and personal causation

In this area of the Resource, the computer is being used as a tool for theory formation. The goal is to develop a theory of how people arrive at an interpretation of the social actions of other people; and to test the theory in certain communication situations, such as the psychiatric interview. The theory is being formulated by Dr. Charles Schmidt, a social psychologist, in collaboration with computer scientists in the Resource. It extends Heider's work on social perception (8) and it employs certain ideas and techniques from work in artificial intelligence.

The specific system that is being developed takes as an input a social episode, i.e., an account of a sequence of actions involving the interactions of several persons within some social context, and it generates an interpretation of the episode in terms of intentions and reasons that might have motivated the persons in the episode to perform their actions.

An episode is presented to the system in the form of a sequence,

$$[S_0, A_1. (S_1), A_2, (S_2), \ldots, A_i, (S_i), \ldots].$$

where S_0 denotes the initial situation; A_1, A_2, A_i denote actions; and S_1, S_2, S_i denote intermediate situations in the episode. Intermediate situations may or may not be present in the description of the episode (the parentheses indicate that this information is optionally present).

In addition to the episode, the system receives as an input a list of beliefs about the entities—persons and objects—that are involved in the episode. A person is described in terms of properties, such as biological states, emotional states, abilities, possessions, knowledges, beliefs, interpersonal relationships, dispositions and wants.

The theory is formulated for the computer in the form of a body of rules of belief and a strategy of interpretation.

The body of rules consists of four parts: *i*) a 'general' set of beliefs about the entities of the social world, *ii*) a set of beliefs about the rules that govern actions in this world (the rules of action), *iii*) a set of beliefs about rules of human motivation (the TRY rules), and *iv*) a set of metarules about personality organization (the Consistency rules).

The rules of action and the rules of motivation play a key role in the process of interpretation.

The rules of action are represented as action schemas in the form,

<Action Name> $(E_1, E_2, \ldots)$;
$PC \rightarrow OC$; <Try Flag>

where $PC = NOC \cup POC$.

In this schema, following the Action Name (e.g., "Give"), the E's are arguments which restrict the class of entities that can be substituted as actor, recipient, object and instrument of the act; PC is a set of preconditions on the act; OC designates the set of outcome conditions associated with the act, and it consists of the set of necessary outcomes (NOC) and the set of possible outcomes (POC), if the act is performed; the <Try Flag> is a binary variable whose value indicates whether the action implies intention on the part of the actor.

Motivational rules (TRY rules) are formulated in the form of a production system which defines (explains) *patterns of actions* in terms of general behavioral concepts, such as hedonism, extended hedonism, reciprocity, and so on.

To illustrate the form of these rules, we present next two simple productions in the system:

$$\text{Hedonism} \rightarrow \text{Immediate} \vee \text{Possible} \vee \text{Self-Enabling}_i \vee \text{Other-Enabling}_i$$

$$\text{Immediate} \rightarrow [\text{Get}\ (A, X) \in HOC_i \wedge V\ (X_A) = <+>] \vee [\text{Lose}\ (A, X) \in HOC_i \wedge V\ (X_A) = <->],$$

where A denotes a person, X a particular object, i is the ordinal position of the act in question, Get (A, X) stands for the act of A getting X, HOC_i represents the set of outcomes of act i that are known to have happened, and $V\ (X_A)$ refers to the 'value' of X with respect to person A.

The production system that represents the TRY rules is quite complex. Some of the rules are defined recursively; in this way a person can be located as a rather remote causal source of a sequence of actions. A *parse* of a connected action sequence in accordance with the TRY rules is a possible behavioral interpretation of the sequence, and thus it is a candidate for an output by the system.

The strategy of interpretation is based on a *postdictive axiom of personal causation* for the 'naive observer', which may be written as,

$$\text{Cause}\ (P, W) \supset \text{Can}\ (P, W) \wedge \text{Try}\ (P, W),$$

where P stands for a person and W for an act or outcome. This statement asserts that if the 'naive observer' believes that a person intentionally caused some act or outcome, then the observer will infer that the person was able to (*can*) do the act or cause the outcome *and* was *motivated* to do the act or cause the outcome.

The following is an outline of the interpretation process.

1)In response to the input presentation of an episode, the system attempts to build a graph of action dependencies. This is done by working backwards from the last action, and searching to determine whether any of the preconditions for this action were created by some previous action. If so, then a link is created connecting the two actions. If no prior enabling act is found, then it is assumed that the precondition existed at the start of the episode and a link is created pointing back to the initial situation. This process is guided by the rules of action, the general set of beliefs about entities in the social environment of the episode, and the beliefs about the specific persons and objects in the episode.

2) Connected graphs of actions are analyzed in accordance with the TRY rules. A set of parses is generated as candidate interpretations of the episode. The parser used here is an adaptation of the flexible parser developed by Dr. Fabens in another area of the Resource (6) for the syntactic analysis of language strings. In this adaptation, the parser uses the TRY rules as a grammar for action sequences.

3) Candidate interpretations are checked by the Consistency Rules. Consistent interpretations are produced as outputs. If consistency rules of personality organization are violated, then a search for other interpretations (in terms of less conventional TRY rule parses) is made.

Thus the interpretation process for an episode consists of a search for a connected structure of actions, the formation of hypotheses about motivational patterns that can possibly explain the episode, and the filtering of these hypotheses by rules of personality organization.

In general, the problem of interpretation in the present area is more complex than the problem of interpretation that we have faced in glaucoma diagnosis. While the control of hypothesis formation in the medical diagnosis problem is largely local (by use of notions of strengths of causation), there is more emphasis on global controls (in terms of patterns of TRY rules and consistency rules) in the problem of social causation.

Initial versions of the computer-based model of social causation have been implemented and tested over episodes involving 3-4 people in sequences of about 10 actions. Several extensions of the model are now underway. Some of this work is directed to more powerful strategies of interpretation—where the approach is to narrow the search for hypotheses by a better utilization of global constraints.

The problem of empirically validating this computer-based model is quite difficult, and it requires not only the development of an appropriate experimental methodology, but also the design of computer facilities for flexible modification and testing of components of the model.

CONCLUDING COMMENTS

Processes of interpretation are important components of scientific and clinical activity. Computer systems can be designed to implement these processes—and thus to substantially assist researchers and clinicians in their work. The mechanization of interpretation processes for given domains (as well as of other problem solving processes) is strongly dependent on the ways in which knowledge about these domains is available to the processes. Thus the development of systems for computer-based interpretation is strongly tied to the development of methods for structuring knowledge (as models, systems of constraints or heuristic rules) in computers. Computer systems with capabilities for convenient acquisition and management of knowledge in a given domain, and with facilities for intelligently performing processes of interpretation, planning (i.e., designing a medical treatment) and theory formation, have the potential for significant impact on research and practice in biomedicine. Work in artificial intelligence, in data base systems, and in interactive languages provides the basis for the design and implementation of such systems. Advances in time sharing and in computer networking are opening the way for collaborative developments of such systems and for their shared utilization by groups of investigators across the country. Our experience in the Rutgers Research Resource shows that systems of this type can be built; and they can be used as powerful tools, for research and problem solving in the life sciences.

REFERENCES

1. Amarel, S. Problems of representation in artificial intelligence. In: *Purposive Systems*, edited by H. Von Foerster et al. New York: Spartan, 1969
2. Amarel, S. On the representation of problems and goal directed procedures for computers. In: *Formal Systems and Non-Numerical Problem Solving by Computer*, edited by R. B. Banerji and M.D. Mesarovic. New York: American Elsevier, 1970.
3. Amarel, S. Representations and modeling in problems of program formation. In: *Machine Intelligence 6*, edited by Meltzer and Michie. Edinburgh: Univ of Edinburgh Press, 1971.
4. Amarel, S., and C. Kulikowski. Medical decision making and computer modeling. *Proc. Subconf. Computers in Biomedicine, 5th Hawaii Intern. Conf. Systems Sci.* January 1972.
5. Buchanan, B., and J. Lederberg. The heuristic Dendral program for explaining empirical data. *Proc. Intern. Federation Information Processing Soc.* 1972, p. TA-2-91-100.
6. Fabens, W. J. H. Pedaglot users manual. *CBM-TR 12* New Brunswick N.J.: Rutgers Univ., Dept. of Computer Sci., 1972.
7. Feigenbaum, E. A., B. G. Buchanan and J. Lederberg. On generality and problem solving: A case study using the dendral program. In: *Machine Intelligence 6*, edited by Meltzer and Michie. Edinburgh: Edinburgh Univ. Press, 1970.

8. Heider, F. *The Psychology of Interpersonal Relations.* New York: Wiley, 1958.
9. Kolker, A. E., and J. Hetherington, Jr. *Becker-Shaffer's Diagnosis and Therapy of the Glaucomas* (3rd ed.) St. Louis: Mosby, 1970.
10. Kulikowski, C. A., and S. Chokhani. Process control model for the regulation of intraocular pressure and glaucoma. *CBM-TR-17* New Brunswick, N.J.: Rutgers Univ., Dept. of Computer Sci., 1973.
11. Kulikowski, C. A., and S. Weiss. Computer based models of glaucoma. *CBM-TR-3* New Brunswick, N.J.: Rutgers Univ., Dept. of Computer Sci., 1971.
12. Kulikowski, C. A., and S. Weiss. Medical Consultant Program - GLAUCOMA (MCP-G1). *CBM-TR 5* New Brunswick, N.J.: Rutgers Univ., Dept. of Computer Sci., 1972.
13. Kulikowski, C. A., and S. Weiss. Strategies for test selection in causal network models. *CBM-TR 11* New Brunswick, N. J.: Rutgers Univ., Dept. of Computer Sci., 1972.
14. Kulikowski, C. A., S. Weiss and A. Safir. Glaucoma diagnosis and therapy by computer. *Proc. Ann. Meeting Assoc. Res. Vision Ophthalmol. May 1973.*
15. Schmidt, C. F. A computer simulation model of the social perception of interpersonal episodes. *CBM-TM 11A* New Brunswick, N.J.: Rutgers Univ., Dept. of Computer Sci., 1972.
16. Schmidt, C. F., and J. D'Addamio. A model of the common-sense theory of intention and personal causation. *Proc. 3rd Inter. Joint Conf. Artificial Intelligence.* Stanford, Calif., 1973.
17. Weiss, S. M. A system for model-based computer-aided diagnosis and therapy. *CBM-TR-27.* (Ph.D. Thesis) New Brunswick, N.J.: Rutgers Univ., Dept. of Computer Sci., 1974.

Computer as a research tool in speech-understanding systems research[1]

D. RAJ REDDY
Carnegie-Mellon University
Pittsburgh, Pennsylvania 15213

Speech-Understanding Systems (2), like several other problems that arise in Artificial Intelligence research, are characterized by high data rates, large amounts of data, and the need to perceive, understand and respond in real time. Thus, while it is acceptable to take 3–5 min to respond to a chess move, one expects an immediate response during a conversation. Unlike the conventional scientific applications, where the solution is known and merely needs to be programmed, in this case not only are the solutions not known but even the research methodology for the discovery of a solution was not well understood a few years ago. Thus the computer plays an important role in the scientific discovery for formulation and validation of models that can explain the perceptual phenomena associated with speech understanding. The role of such an interactive computer research system in information processing psychology is analogous to the role of an electron microscope in neurobiology. In this paper we will briefly discuss the structure and organization of a working speech-understanding system and discuss the facilities in our research environment and their relevance to this research.

THE HEARSAY SPEECH-UNDERSTANDING SYSTEM

There are several active research groups working towards systems for speaking to computers.[2] In this paper we will discuss the problems that arise in attempting to speak to computers with specific examples from the Hearsay system being developed at Carnegie-Mellon University.

We all know that a native speaker uses, unconsciously, his knowledge of the language, the environment and the context in understanding a sentence. These sources of knowledge include the characteristics of the sounds (acoustic-phonetics), the stress and intonation patterns of speech (prosodics), a dictionary of legal words (lexicon), the grammatical structure of the language (syntax), the meaning of words and sentences (semantics), and the context of the conversation (pragmatics).

To illustrate the problems of speech recognition by computers let us consider the following sentences.

1. Colorless paper packages crackle loudly.
2. Colorless yellow ideas sleep furiously.
3. Sleep roses dangerously young colorless.
4. Ben burada ne yaptigimi bilmiyorum.

---

[1] This work was supported by the Advanced Research Projects Agency of the office of the Secretary of Defense (F44620-73-C-0074) and is monitored by the Air Force Office of Scientific Research. This document has been approved for public release and sale; its distribution is unlimited.

[2] BBN, CMU, IBM, SDC and XEROX to name a few. See IEEE Transactions on acoustics, speech and signal processing (Dec. 1974 and Feb. 1975) for descriptions of several of these systems.

IN MUD EELS ARE, IN CLAY NONE ARE

in mudies sar in clay nanar
in my deals are en clainanar
in my ders en clain
in model sar in ? anar
in mudilzar in claynanar

Figure 1. Perceptual responses of subjects to the sentence "In mud eels are in clay none are."

IN PINE TAR IS, IN OAK NONE IS

in pine tarrar in oak ? es
in pyntar es in oak nonnus
in pine tar is in ocnonin
in pinatar in ocknanus
en pine tar is in oak is

Figure 2. Perceptual responses of subjects to the sentence "In pine tar is in oak none is."

The first sentence, though grammatical and meaningful, is lacking pragmatically. The second is syntactically correct but meaningless. The third is both syntactically and semantically unacceptable. The fourth is completely unintelligible. One would expect a listener to have more difficulty recognizing a sentence if it is inconsistent with one or more of the above sources of knowledge. In fact Miller and Isard (1) show that this is indeed the case.

How does one use all these sources of knowledge in understanding an utterance? Modern linguistic theory is of little help, because it attempts to handle every possible sentence using primarily the syntactic source of knowledge. Some sentences may have two or more syntactic interpretations. For example the sentence, "I saw the Statue of Liberty flying over New York," leads to a syntactic ambiguity. Who is flying? Is it "I" or is it the "Statue of Liberty"? In a system where many diverse sources of knowledge are actively cooperating, a simple syntactic theory capable of handling a vast majority of sentences may be adequate. Rare cases, where a simple theory fails, can be resolved using other sources of knowledge, such as context.

How is a given source of knowledge used in recognition? The Shannon experiment (8) gives a clue. In this experiment, human subjects demonstrate their ability to predict (and correct) what will appear next, given part of a sentence. We propose that anticipation and subsequent verification are principal mechanisms used in human perception.

This hypothesize-and-test paradigm works in most cases. However, if the knowledge is incomplete or inaccurate, people will tend to make erroneous hypotheses. This can be illustrated by a simple experiment. The subjects were asked to listen to two sentences and write down what they heard. The sentences were "In mud eels are in clay none are," and "In pine tar is in oak none is." The responses of five subjects are given in Figs. 1 and 2. These figures show that the listener forces his own interpretation of what he hears, and not necessarily what may have been intended by the speaker. In Fig. 1, because the subjects do not have the contextual framework to expect the words "mud eels" together, they write more likely sounding combinations such

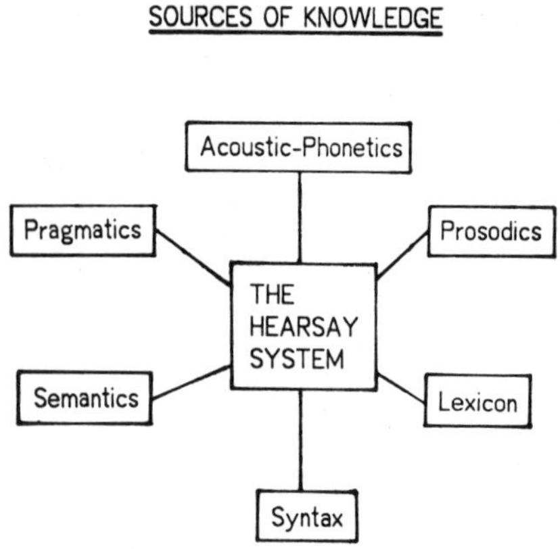

Figure 3. Representation of knowledge in Hearsay as a set of cooperating parallel processes.

as "my deals" or "models." In Fig. 2, we find the same problem with words such as "oak none is." Notice that they failed to detect where one word ends and another begins. It is not uncommon for machine recognition systems to have similar problems with word segmentation.

To equal human performance, a machine must also use all the available sources of knowledge effectively. In the Hearsay system this is achieved by representing knowledge as a set of cooperating parallel processes as shown in Fig. 3. Just as in the experiment, the Hearsay system also uses all the sources of knowledge to generate hypotheses about what word

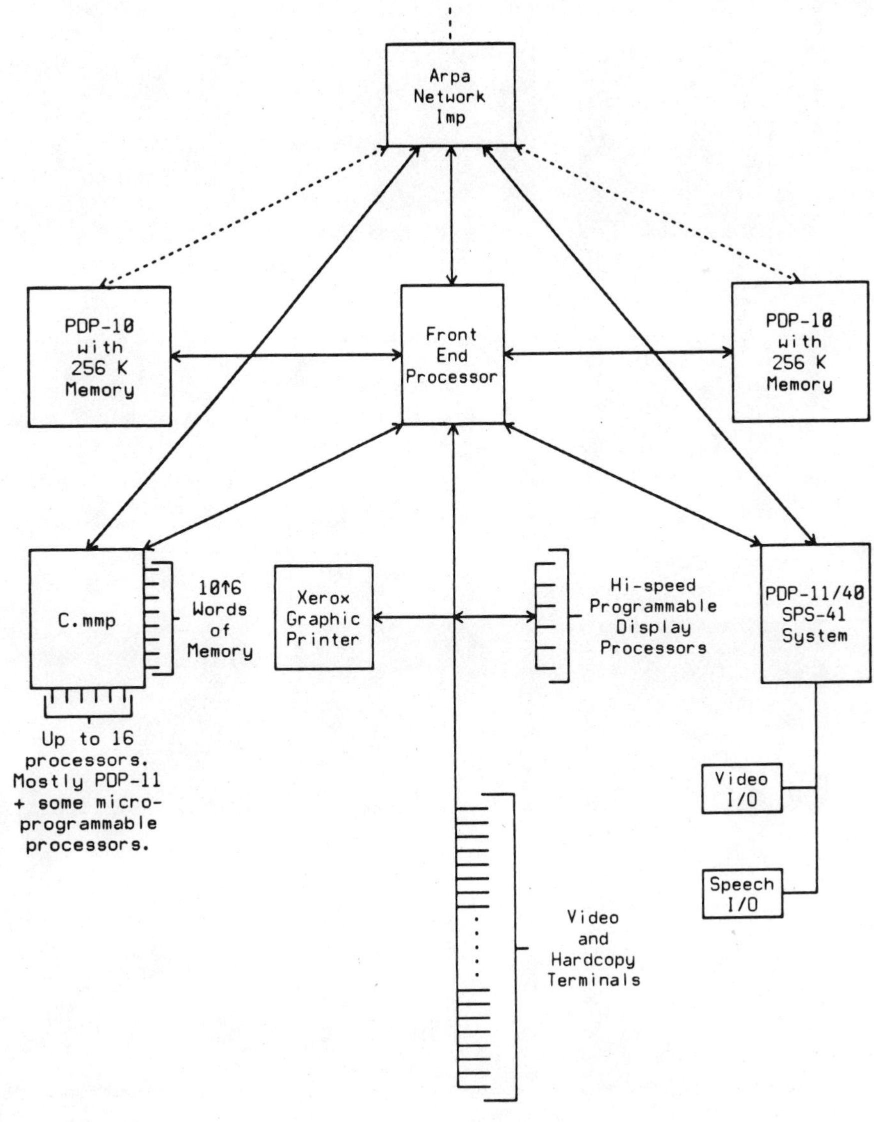

Figure 4. The computer science research facility at Carnegie-Mellon University.

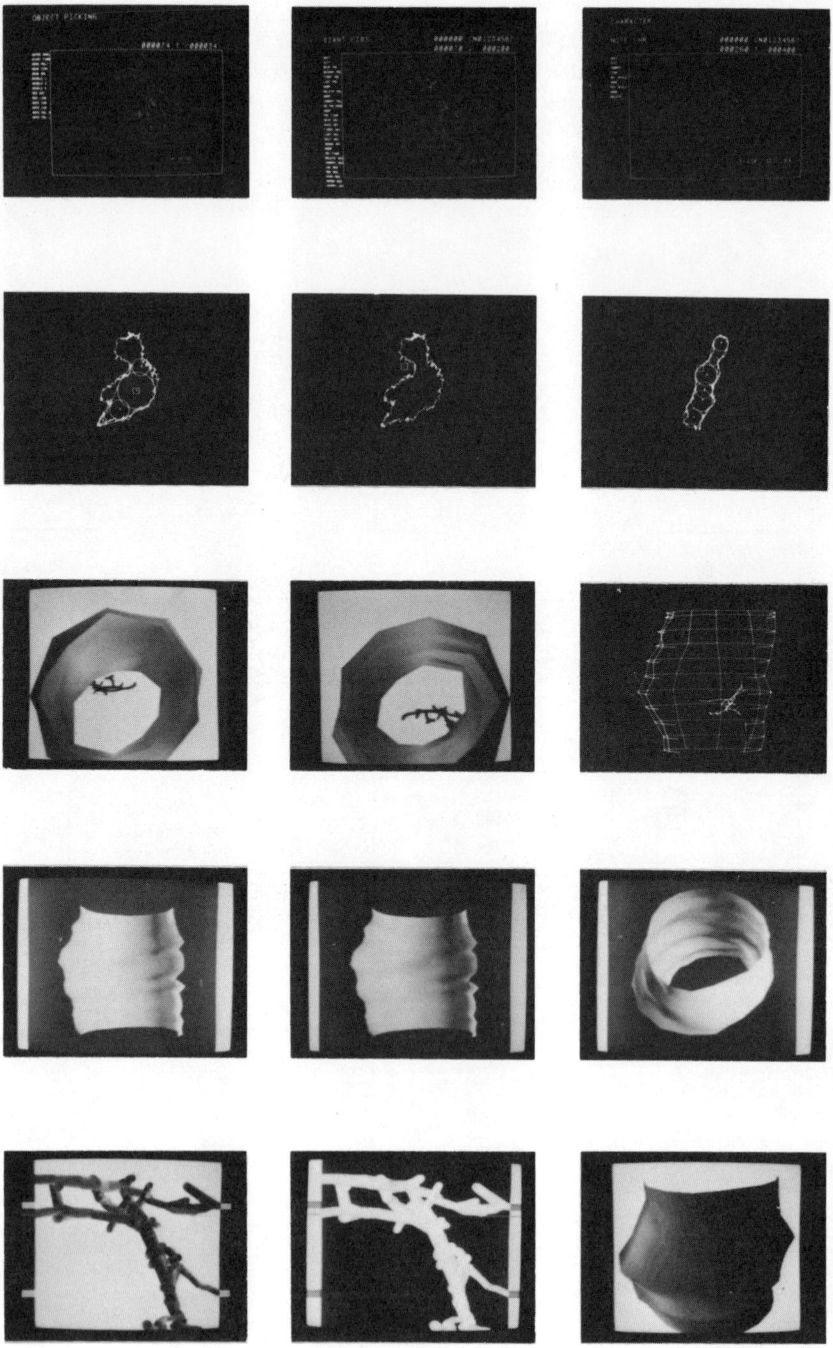

Figure 5. Typical examples of output using vector and video graphic facilities.

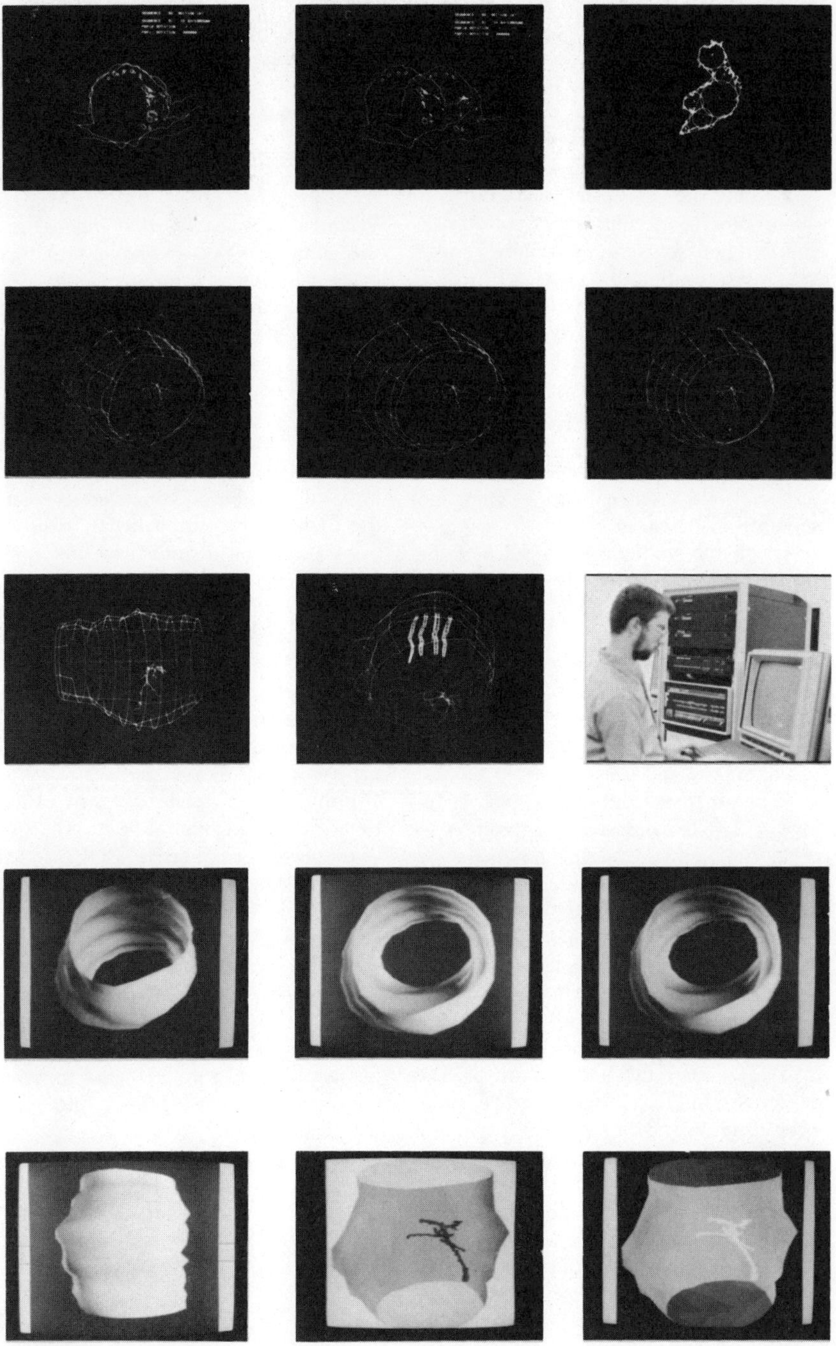

might appear in a given context, or to reject a guess. When the system makes errors it is usually because the present state of its knowledge is incomplete and possibly inaccurate. The model, the system and the tasks that Hearsay works on have been described elsewhere (5–7).[3]

To date the system has been tested on about 144 connected speech utterances, containing 676 words, spoken by 5 speakers, and consisting of 4 tasks, with vocabularies ranging from 28 to 76 words. It takes about 5–10 sec to recognize an utterance. On the average the system is able to locate and identify about 89% of the words correctly using all the sources of knowledge; without semantics this accuracy decreases to 67%. It decreases further to 44% when neither syntax nor semantics is used.

To be acceptable as an input medium, the system must not only perform with high accuracy, but it must also minimize the combinatorial explosion so as to respond as fast as a human would in a similar situation. To achieve this it is our belief that the system performance will have to improve to about 90% without the use of syntax and semantics.

As we increase the size of the vocabulary to 1,000 words or more and as we relax the structure of the language, the need for effective use of all the sources of knowledge to control the combinatorial explosion becomes increasingly critical if we are to achieve the objective of approaching human performance in speech perception.

Our preliminary studies (3) indicate that to achieve the high accuracies in a large vocabulary system performing in real time, we will need computational power of about 100 mips (million instructions per second). Since commercially available processors are about 10 to 100 times slower, it becomes necessary for us to consider multiple processor machine architectures eventually capable of delivering the 100 mips computational power. The C.mmp, a multiminiprocessor computer (9) system developed at Carnegie-Mellon University, is one such. In the rest of this paper we will discuss the research computer facilities available in our laboratory.

THE RESEARCH COMPUTER ENVIRONMENT

We have stated earlier that we have to use the computer as a research tool to discover satisfactory solutions to the speech-understanding problem. To be useful in this role, the computer must satisfy several requirements: it must be interactive, it must be capable of handling vast amounts of raw data and extensive processing, it must be possible to conveniently add high data rate real time devices, it must be possible to conveniently observe the dynamic execution behavior of large programs, and it must be possible to obtain hardcopy graphical output for off-line examination data analysis. Figure 4 gives the computer facility at Carnegie-Mellon University used for computer science research.

The large computational power requirement is satisfied by two PDP-10 processors with 256K memory each, the C.mmp multiprocessor system with functionally specialized and microprogrammable processors, and a PDP-11 based SPS-41 signal processing system. The C.mmp can have up to 16 processors with some processors capable of 5 to 10 mips. The SPS-41 can do a 1024 point FFT in 8 msec (or approximately 15 mips). All these systems are available on an interactive basis from any terminal within the facility. These are connected to various processors using a PDP-11 line concentrator front-end.

The high data rate speech (200 thousand bits per sec) and image (100 million bits per sec) devices are connected to the PDP-11/SPS-41 system which is the

[3] A movie illustrating the performance and operation of the system can be obtained by writing to the author.

main signal processing facility within the system. The data can be transfered among various systems within the facility using the line concentrator, ARPA network, diskpacks, magnetic tape, or DEC tape.

The primary research terminal for speech, vision, and neuronal reconstruction research is the high performance programmable display processor developed at Carnegie-Mellon University. Each terminal consists of a 16K PDP-11/15, a special display processor and an HP-1310A electrostatic display. It can display about 6,000 characters or about 50,000 short vectors flicker free and features soft characters, digital vector generalization, real-time 3-D rotation and 10ns 10-bit D-A converters. These displays are essential for a more meaningful graphical representation of the large amount of data and the dynamic execution behavior of complex processing tasks. Also available is a high band-width color video display which is refreshed from the random-access memory. Figure 5 gives some typical images produced using these displays in our research on neuronal reconstruction of histological sections (4).

The XGP (Xerox Graphic Printer) is a hardcopy graphical output device that can produce arbitrary computer output of variable font text, vectors, and half-tone images at 10 sec a page. Figure 6 gives an example of XGP output from our reconstruction research. The structure of various graphics systems developed at our laboratory is described by Reddy, Broadley, Kriz, Powell and Rosen (to be published).

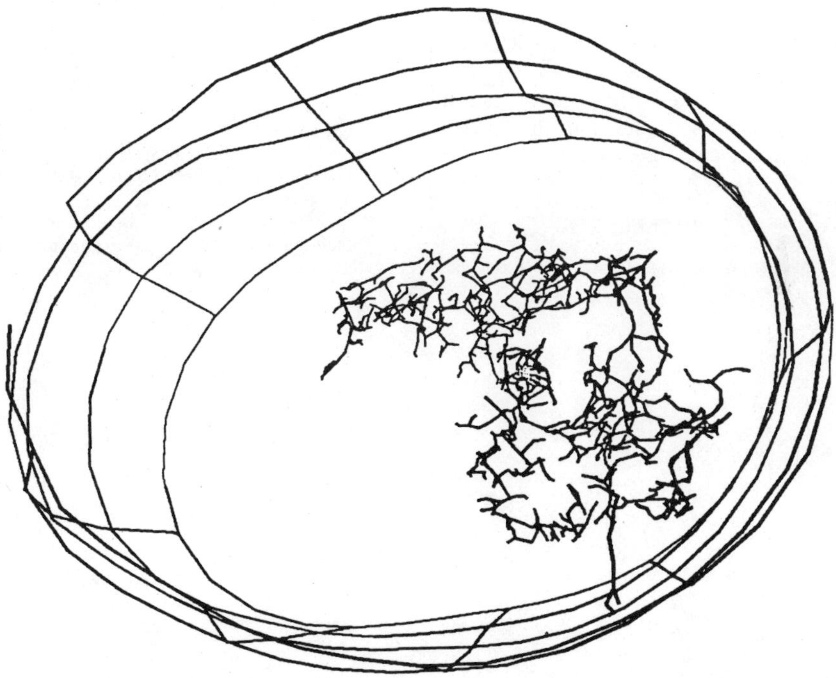

Figure 6. Typical example of hardcopy graphic output using the Xerox graphic printer.

CONCLUSION

We have illustrated a research application of computers, i.e., the speech-understanding research, which pushes the frontiers of computer science and technology. Though the present cost of our research computer system is over a million dollars, we expect that by the early 1980's systems of comparable power and sophistication will be available for under a hundred thousand dollars. At that time, any research group which can afford a mass spectrometer or an electron microscope will be able to afford this type research computer system for use in their own laboratory.

The main contribution of this research to other sciences is the novel way in which it is able to use diverse sources of knowledge such as pragmatics, semantics, syntax, phonology, and acoustic-phonetics in the decoding of the utterance. These knowledge sources are organized into a set of independent, and yet cooperating, parallel processes that predict what may appear in a given context or to verify, reject, or reorder a hypothesis resulting from a previous prediction (the hypothesize-and-test paradigm).

REFERENCES

1. Miller, G. A., and S. Isard. Some perceptual consequences of linguistic rules. *J. Verbal Learning Behav.* 2: 217–228, 1963.
2. Newell, A., J. Barnett, J. Forgie, C. Green, D. Klatt, J. C. R. Licklider, J. Munson, R. Reddy and W. Woods. Final report of a study group on speech understanding systems. (First published in report form, 1971). Amsterdam: North-Holland, 1973.
3. Reddy, D. R. Some numerical problems in artificial intelligence: implications for complexity and machine architecture. In: *Complexity of Sequential and Parallel Numerical Algorithms,* edited by J. F. Traub, New York: Academic, 1973, p. 131–147.
4. Reddy, D. R., W. J. Davis, R. B. Ohlander and D. Bihary. Computer analysis of neuronal structure. In: *Intracellular Staining Techniques In Neurobiology,* edited by S. B. Kater and C. Nicholson. New York: Springer-Verlag, 1973, 227–254.
5. Reddy, D. R., L. D. Erman, R. D. Fennell and R. B. Neely. The Hearsay speech understanding system: an example of the recognition process. *3rd. Intern. Joint Conf. Artificial Intelligence.* Stanford, Calif. 1973, p. 185–193.
6. Reddy, D. R., L. D. Erman and R. B. Neely. A model and a system for machine recognition of speech. *IEEE Trans. Audio Electroacoustics,* AU-21 (3): 229–238, 1973.
7. Reddy, D. R., and A. Newell. Knowledge and its representation in a speech understanding system. In: *Knowledge and Cognition,* edited by L. W. Gregg, Washington, D. C.: Lawrence Erlbaum Assoc., In press.
8. Shannon, D. E. Prediction and entropy of printed English. *Bell System Tech. J.* 30: 50–64, 1951.
9. Wulf, W. A., and C. G. Bell. C.mmp: A multi-mini-processor. *Proc. Fall Joint Computer Conf.* 1972, p. 765–778.

Computer-based auditory neurophysiology laboratory[1]

J. E. HIND[2] AND W. S. RHODE[3]
University of Wisconsin Medical School
Madison, Wisconsin 53706

For nearly 2 decades our laboratory has carried on microelectrode studies of the auditory nervous system. A primary objective of these studies is to determine how the properties of an acoustical stimulus are represented or coded in the temporal patterns of all-or-none spike discharges that may be recorded from single neural units (cell bodies or fibers). For many years, following the pioneering observations of Galambos and Davis (3), the only available methods for study of unit discharge patterns involved either laborious manual measurement of oscilloscope tracings or rough estimation of the threshold of unit discharge by listening to the amplified activity. Although qualitative techniques were developed to deal with the massive amounts of data that result from experiments of this kind, it was not until Gerstein and Kiang (4) and their colleagues at MIT introduced the digital computer that it became possible to perform precise statistical analyses of unit discharge patterns as they occur in real time.

Our initial utilization of digital computers for spike train analysis occurred in 1961 and employed a CDC 160 computer in the University of Wisconsin Computing Center to measure the times of occurrence of unit spikes reproduced by playback of analog magnetic tape recordings (5). However, since we had access to the 160 only a few hours each week and the results were not available while an experiment was in progress, the full benefits of computer utilization were not realized until 1963 when a LINC (Laboratory Instrument Computer) was installed in our laboratory under the LINC Evaluation Program (2). This report will describe how the LINC has been integrated into our research program and is intended to illustrate how a relatively simple, nonmathematical application of computers can enormously expedite experimental progress in one field of life sciences research.

DESCRIPTION OF LINC-BASED LABORATORY

Figure 1 is a simplified block diagram of the major components in our auditory laboratory. The most essential element in the interface between the LINC and the experimental apparatus is a custom-built digital event timer that includes a binary counter and buffer register to measure time intervals between data events by registering the number of pulses from a time base generator. Under program con-

[1] This research was aided by NIH grants NS 06225 and RR 00249.

[2] Organizer of this Session of the Computer Conference.

[3] Also Director of the Laboratory Computer Facility.

trol, any of several data signals can be selected to cause storage and/or resetting of the counter; these data signals include pulses coincident with unit spikes, with the onset of stimulus tones, and with the positive-going axis crossings of a sinusoidal stimulus waveform. The event timer is a very busy element in an experiment, one moment measuring the period of the stimulus oscillator, the next instant registering the time of occurrence of unit spikes relative to the phase of the sinusoidal stimulus.

By means of a stepping motor attached to the frequency dial of a General Radio 1309A oscillator, LINC output pulses control the stimulus frequency which can be set to within 2 parts per 1,000. Similarly, LINC output pulses control the presentation of stimulus tones, and their intensity is modulated in 1 decibel (dB) steps by a custom-built, relay-operated attenuator. At the beginning of each experiment the LINC records the sound pressure level (SPL) at all frequencies of interest by means of a calibrated probe tube microphone. Through use of a logarithmic converter (Pacific Measurements Corp. model 1020), microphone output is sampled by the LINC analog/digital converter and expressed as dB SPL over an 80 dB dynamic range.

When a single unit has been isolated by the microelectrode, the experimenter specifies interactively via the LINC display and keyboard the desired range of stimulus frequencies and sound pressure levels. The LINC then activates the stimulus presentation system, sets the stimulus parameters, and collects and analyzes the neural data. Immediately following completion of a stimulus sequence, the results are presented on the LINC display to guide the next choice of stimuli, a factor of crucial importance in maximizing the efficiency of the experimental strategy. It is now common for recordings to be maintained from a neuron for periods in excess of 10 hours, during which several thousand stimuli may be presented and the times of occurrence of up to a million spikes determined with a

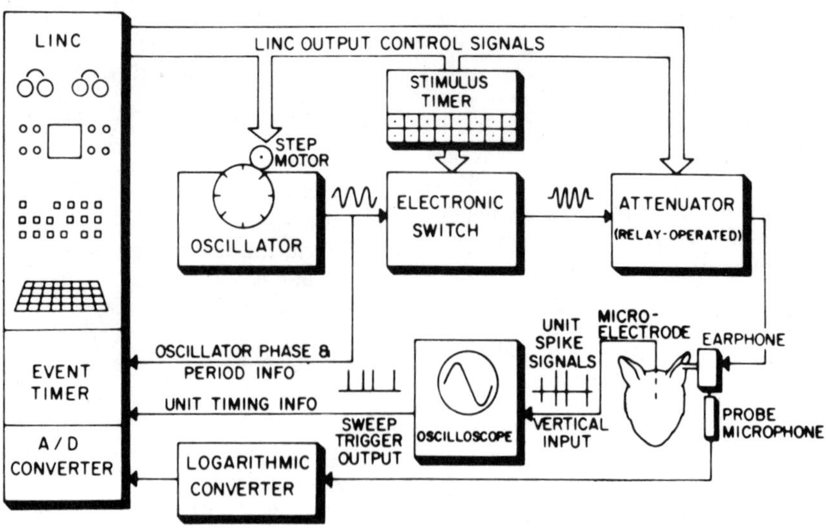

Figure 1. Block diagram of LINC and major apparatus in auditory neurophysiology laboratory.

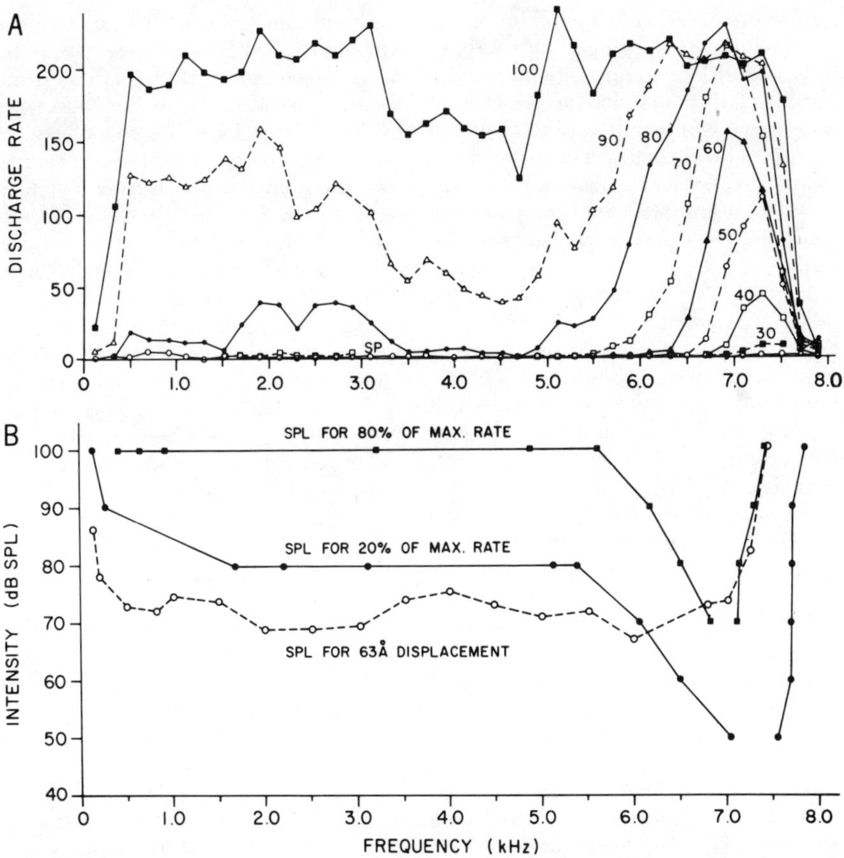

Figure 2. *A*) response area for a squirrel monkey auditory nerve fiber: discharge rate versus frequency with sound pressure level as a parameter. *B*) solid curves: iso-rate functions replotted from data in *A* showing SPL required to produce 20% and 80% of maximum discharge rate; dashed curve: SPL required to produce a constant 63 Å displacement of the basilar membrane as measured with the Mössbauer technique.

resolution as fine as 2 μsec. The volume and precision of the data made possible by computer utilization have revolutionized experimental auditory neurophysiology.

EXAMPLES OF EXPERIMENTAL RESULTS

Figure 2*A* presents the response area for an individual auditory nerve fiber recorded in the anesthetized squirrel monkey. Each curve shows the discharge rate in spikes per second in response to a series of pure tones at the specified sound pressure level as the frequency is varied in discrete steps. Note that at 40 dB SPL only frequencies near 7.3 kHz were effective in eliciting a discharge, whereas at 100 dB the responsive range broadened to considerably lower frequencies. In contrast to the traditional "tuning curve" for an auditory unit which merely plots threshold SPL as a function of frequency and thus fails to describe the information transmitted by the unit at suprathreshold intensities, the computer-automated response area provides a quantitative description of unit response throughout the entire effective frequency/intensity domain.

In order to understand the coding performed in the cochlea, the transformation of acoustic energy into neural signals must be explained. The mechanics of this transformation occur largely in the organ of Corti which contains the sensory hair cells or transducers. The organ of Corti rests on the basilar membrane which appears to be the primary determinant of the mechanical motion of the organ. In order to explain the transduction, the vibratory characteristics of the basilar membrane need to be known. One of us (6) has studied its vibration utilizing the Mössbauer effect which is a Doppler phenomenon at the nuclear level that permits the measurement of very small velocities, of the order of 0.1 mm/sec. A small radioactive source, 5 μm × 60 μm × 60 μm, is placed on the vibrating structure. Gamma-rays emitted by the source pass through an absorber; the number absorbed is a function of the relative velocity between the source and absorber. This modulation of gamma-ray activity provides the basis for determining the velocity waveform of the vibrating structure. The LINC is used to collect, display and analyze the gamma-ray data. The analysis occurs off-line due to inadequate computing capability of the LINC. The result of these experiments is a plot of amplitude and phase of vibration versus frequency. The vibratory data can then be compared to the comparable neural data for an auditory nerve fiber. The neural data shown as isointensity contours in Fig. 2*A* are replotted in Fig. 2*B* as isorate contours which depict the SPL required to produce 20% and 80%, respectively, of the maximum observed discharge rate. The dashed curve in Fig. 2*B* shows the SPL required for a constant 63 Å displacement of the basilar membrane as measured with the Mössbauer method. The difference in shape between the mechanical and neural tuning characteristics is obvious; much more investigation will be required before the factors that link the mechanical vibration and neural excitation are clearly understood.

In addition to determining the average discharge rate as displayed in the response area, the computer offers a powerful tool for study of the fine time structure of the discharges. It has long been known that auditory nerve fibers exhibit phase locking which is the tendency for the discharges to occur in a restricted segment of the cycle of a periodic, low frequency stimulus. This results in synchronous volleys of impulses in the array of auditory nerve fibers, with the rhythm of the volleys tending to reproduce the stimulus periodicities. Although there is no direct evidence that the central nervous system utilizes the temporal rhythm of the neural volleys in determining the perceived pitch of a stimulus, the phase-locking effect provides a useful analytical means to study the cochlear transduction process. We have utilized complex periodic stimuli composed of two harmonically related pure tone components (1). The computer times the occurrence of spikes with respect to the fundamental frequency of the complex wave and produces a period histogram in which the length of the abscissa equals the period of the complex wave and the contents of the histogram bins indicate the distribution of spikes throughout the period. Figure 3 presents period histograms obtained from a squirrel monkey auditory nerve fiber in response to a complex tone composed of 1,000 and 1,500 Hz components at fixed intensities of 70 and 60 dB SPL, respectively, as the phase angle between the components is varied in the acoustic stimulus. The histograms illustrate the distribution of spikes throughout the period of the complex wave; superimposed on the histograms is a smooth curve which was computer-generated by addition of two sinusoids of corresponding frequency. In Fig. 3*A* the amplitudes and phase angles of the sinusoids were adjusted to yield a least-squares fit to the histogram; in the other parts of the figure the amplitudes determined in Fig. 3*A* were retained, but the phase shifts introduced in the acoustic stimuli were reproduced in the computer-generated curves. It can be seen that the actual distribution of spikes is in good accord with the computer-generated curves, demonstrating that the cochlear transduction process is linear for phase under these circumstances. We have used the two-tone technique extensively, varying the amplitudes and phase angles of the two components widely. In general, there is a nonlinear interaction between the two components; the precise nature of the nonlinearity remains the object of intense study both in our laboratory and elsewhere.

NEW DATACRAFT COMPUTER SYSTEM

In order to pursue more complicated experiments, gather larger amounts of

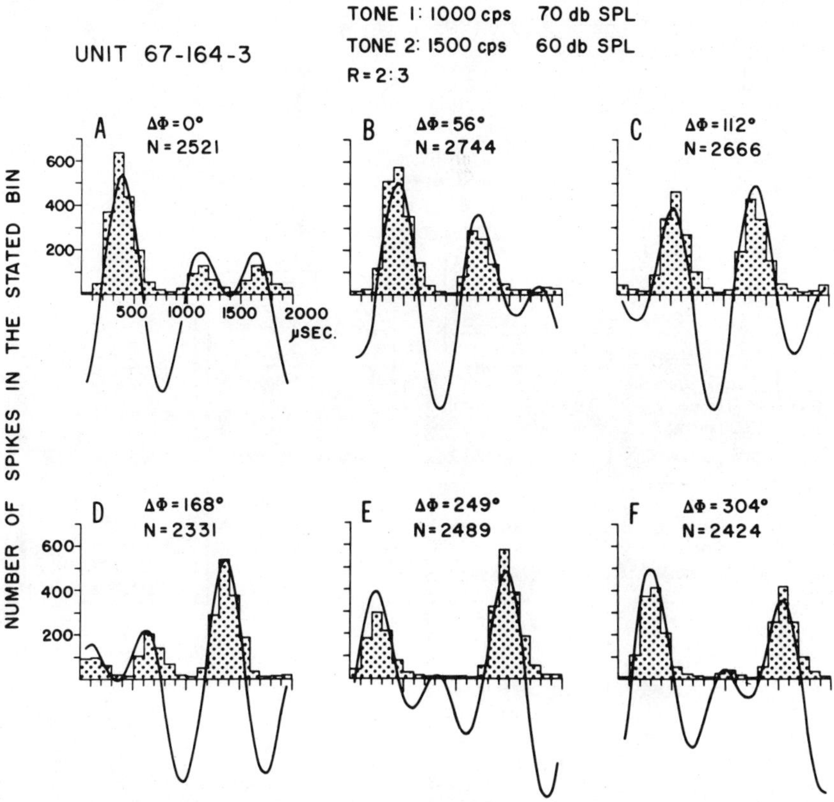

Figure 3. Period histograms for squirrel monkey auditory nerve fiber discharges in response to complex periodic stimulus consisting of 1,000 and 1,500 Hz components with the indicated phase relations. Smooth curves superimposed on histograms are computer-generated by addition of sinusoids of appropriate frequencies with specified phase relations.

data and have greater computing capabilities available, both on-line and off-line, we have recently acquired a Datacraft 6024/5 computer with 24-K words of 24-bit, 0.95 μsec core memory and a 10.8 Mbyte disc drive. This system will be integrated into laboratory operations as the various components shown in Fig. 4 are completed. The new computer permits us to program in a higher-level language, Fortran, as well as in assembly language, and it has an elegant disc monitor operating system which controls all the programming systems and support modules such as Fortran, RPGII, Basic, Forgo (a diagnostic Fortran load and go system), Snobol, Acronim (a reentrant interactive editor), and an overlaying LINK cataloger along with a file manager. Multiple programs can be simultaneously executing at various priorities while programs are compiled at the lowest priority in the system. Previously we had to develop our own ad hoc operating system for each experiment. While this permitted optimization of the use of the system it led to many specific developments that were often not usable by anyone else. Furthermore, assembly language programming for experimental control has to be coupled with a reasonable knowledge of the equipment to be controlled, a task of sufficient difficulty that few of our colleagues really mastered

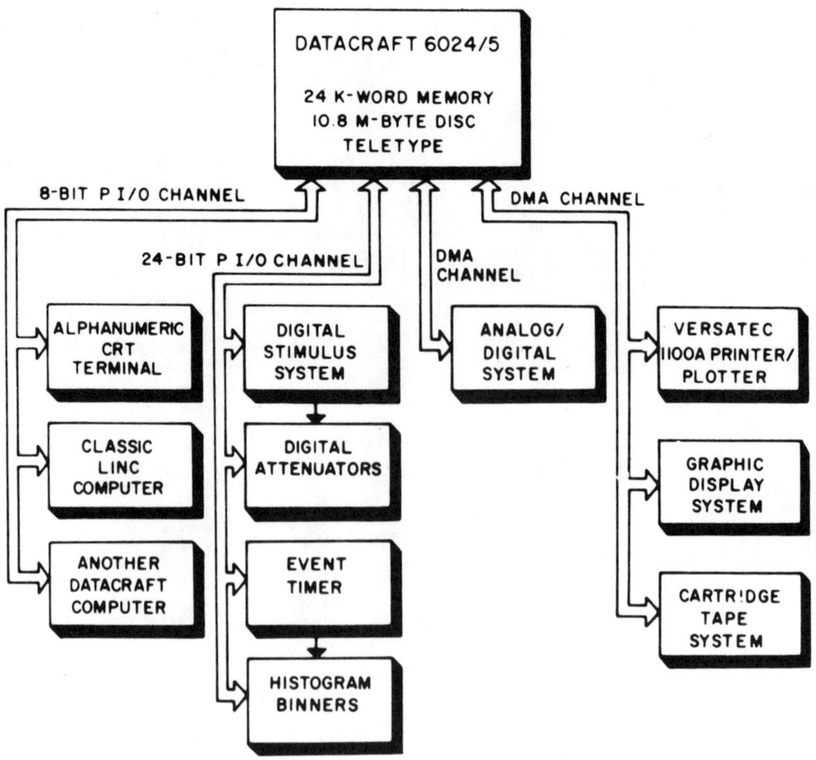

Figure 4. Planned configuration for Datacraft computer system including special equipment for auditory research and connections to classic LINC and another Datacraft in a central facility.

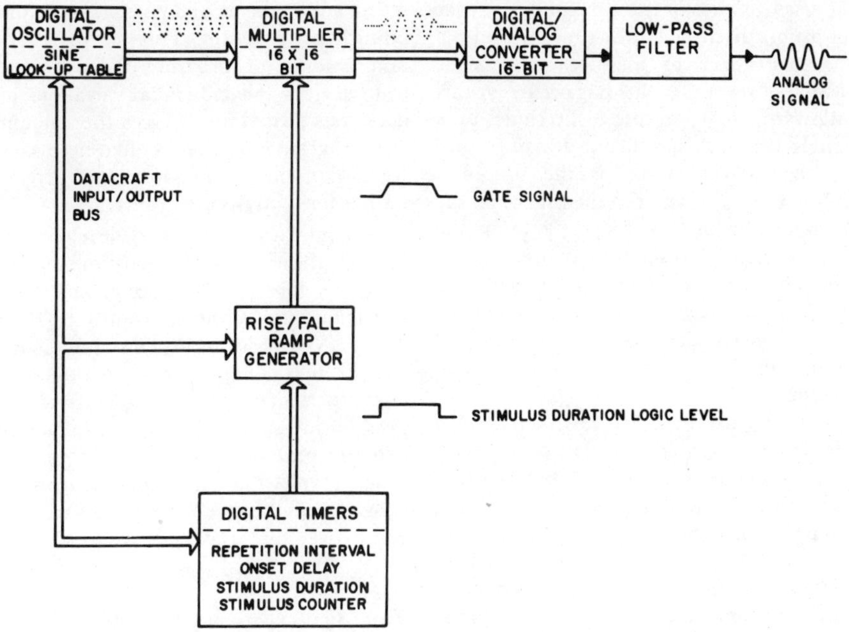

Figure 5. One channel of a 3-channel digital stimulus system consisting of a digital oscillator, digital electronic switch, and a set of digital timers with a stimulus counter. The computer input/output bus controls the system components. The digital oscillator functions by extracting the value of the sine function from a table for a specified phase angle; the rate at which the phase is changed determines the frequency. The combination of the rise-fall generator and digital multiplier constitutes an electronic switch, a device which reduces the effect of transients by turning the stimulating waveform on and off gradually; the waveforms which are multiplied are a sinusoid and a trapezoid. The output of the multiplier feeds a digital-to-analog converter, the output of which is filtered to remove high frequency noise and then attenuated to drive the earphone which produces the acoustic stimulus.

it. While it will continue to be necessary to write assembly language handlers to control hardware devices, these handlers can be called in Fortran programs which will hopefully make it easier for investigators to create and modify their programs.

The new system involves extensive hardware developments, including a custom-designed digital stimulus generator, a block diagram of which is shown in Fig. 5. The system consists of a digital oscillator, digital electronic switch, and a set of digital timers with a stimulus counter. The output waveform can be a sine, triangle, square, or sawtooth wave, and the precision with which the frequency and phase can be controlled as well as the freedom from distortion are expected to substantially exceed the level of performance that can be achieved with analog circuitry; a stimulus with lowest possible distortion is essential for certain studies of the nonlinearities inherent in the ear.

Another important hardware development is a multichannel event timer which is basically a 24-bit counter with a programmable time base of 1–999 μsec. The use of a 24-bit register allows times of 8 sec to be measured without overflow with 1 μsec resolution. A single word can then be used to encode the time of occurrence of a unit discharge. The timer includes a status register to identify and encode which of several events has occurred; this allows up to 16 events to be timed by a single timer. A significant design feature of the interval timer is the use of a 32-word buffer between the timer and the computer which will greatly relax the requirements for short response time on the part of the computer. Since there will be many devices and several individuals using the computer simultaneously, the ability to ensure against missed data will be important.

The hardware effort also includes the design of a graphic display device with 512 × 512 resolution capable of writing one point each microsecond and being completely buffered. This display uses standard TV technology by piggybacking on an alphanumeric terminal. We therefore combine several attractive features: relatively low cost graphics, fast writing speed permitting dynamic displays, and a fully buffered display which minimizes CPU support of the terminal.

Besides the heavy commitment to hardware design and development which has been a continual one since the introduction of the LINC computer, a rather extensive effort in the software area is under way, beginning with the creation of a general-purpose response area program which incorporates all of the varieties of analysis we have found useful in dealing with the temporal discharge properties of units. This program was written within the context of a data base management system which includes a schema and a subschema. The schema is a description of a data file and includes information that determines its structure and specifies the data item names and type (integer, real, string), and whether it occurs more than once. There is a separate schema for each experimental paradigm. There is also a separate subschema or description of the data items and types for each application program. A monitor program interprets the specification in the subschema and schema and passes data to the application program. This allows a certain amount of program independence from devices and individual variations in data structure. This reduces the amount of reprogramming that is often necessary when data files are not in the proper format for analysis programs.

One final area of software development involves the creation of a real-time control language for use in experimental data collection and stimulus control. While our LINC programs allow interactive specification of experimental parameters, our experience has convinced us of the need for investigators to be able to modify experimental paradigms quickly, sometimes even while an experiment is in progress, and we are working toward that end.

THE COSTS

From our experience with the LINC it seems reasonable to assume a useful life of at least 10 years for a laboratory computer system of this kind. Amortized over 10 years, the average yearly capital expenditure for the LINC system including peripherals has been around $6,000; the corresponding figure for the Datacraft will probably be about $8,000. By contracting for maintenance service from our institution's Medical Electronics Laboratory, our yearly costs are $850 and $1,700 for the LINC and Datacraft, respectively, but these figures would be substantially higher if maintenance services were purchased from the manufacturers. In addition to maintenance, we have enjoyed the support of a half-time design engineer throughout the period since the LINC was installed; we consider such engineering service absolutely essential for maximum exploitation of a computer in a laboratory setting in order to meet the ever-changing needs for interfacing of laboratory apparatus and the creation of specialized equipment that simply cannot be purchased. Our department has also employed a full-time programmer throughout this period, but most of our programs have been written by the research investigators who use the computers. Thus anyone contemplating a similar operation should anticipate that the costs for engineering and programming will far exceed the capital expenditures, perhaps by a factor of from two to five times.

The yearly costs for consumable supplies for the LINC and the Datacraft are very modest indeed in comparison with the capital and personnel expenditures, currently totaling about $4,000 for both machines.

REFERENCES

1. Brugge, J. F., D. J. Anderson, J. E. Hind and J. E. Rose. Time structure of discharges in single auditory nerve fibers of the squirrel monkey in response to complex periodic sounds. *J. Neurophysiol.* 32: 386–401, 1969.
2. Clark, W. A., and C. E. Molnar. A description of the LINC. In: *Computers in biomedical research,* edited by R. W. Stacy and B. Waxman, New York: Academic, 1965.
3. Galambos, R., and H. Davis. The response of single auditory-nerve fibers to acoustic stimulation. *J. Neurophysiol.* 6: 39–57, 1943.
4. Gerstein, G. L., and N. Y.-S. Kiang. An approach to the quantitative analysis of electrophysiological data from single neurons. *Biophys. J.* 1: 15–28, 1960.
5. Hind, J. E., and H. Ludwig. Adaptation of the CDC 160 computer to measurement of the temporal discharge characteristics of single neural units. *Proc. 16th Ann. Conf. Eng. Med. Biol.* 5: 64–65, 1963.
6. Rhode, W. S. Observations of the vibration of the basilar membrane in squirrel monkeys using the Mössbauer technique. *J. Acoust. Soc. Am.* 49: 1218–1231, 1971.

U.C.W. ABERYSTWYTH LIBRARY

Session II—Computer applications II
Rapporteur's summary

BERNARD SALTZBERG
Tulane University School of Medicine
New Orleans, Louisiana 70112

The five papers in Session II on computer applications cover a range of scientific problems which emphasize the essential role of modern computer technology and in particular interactive computer graphics in conducting advanced research in molecular biology, clinical medicine, and the neural sciences. The work of Bob Langridge in the application of interactive computer graphics to the study of large biological molecules, and the CARTOS system described by Cy Levinthal to produce three-dimensional reconstruction from serial histological sections to expand our knowledge of the anatomical relationships among nerves, are representative of the role that computers are playing in expanding the horizons of basic biological research. The insightful papers by Saul Amarel and D. R. Reddy, as well as the remarks of the session chairman, Ed Feigenbaum, represent exciting introductions to the application of artificial intelligence (AI) to important problems in the life sciences. The paper by Amarel introduces the reader to the applications of AI to medicine and psychology, while the paper by Reddy describes the application of AI to computer speech understanding. The fascinating neural science research described by Joe Hind demonstrates the vital role of computers in research in auditory neurophysiology, especially in problems of spike train analysis arising in studies of auditory stimulus coding.

The discussions of these papers, during and after the formal presentations, suggested that the non-computer users at the meeting were still seeking closure in terms of how to identify problems in the life sciences in which collaboration of life scientists and computer scientists would enhance research. The attendees rightly observed that most of the papers presented at the meeting represented research in the life sciences conducted by an investigator who was also a fairly skilled computer scientist. They commented that in such circumstances, it is not difficult to understand the mechanisms by which computers are successfully merged into the solution of problems in the life sciences. A number of attendees contended that the problem needing attention is that of defining mechanisms and procedures whereby life scientists with limited computer knowledge (people with problems looking for solutions) can be successfully wedded to computer scientists (people with solutions looking for problems).

The successful merging of these two groups is a challenge to the research community. In the past, many ad hoc combinations of life scientists have produced disappointing results. Bad mergers seriously compromise both the biologist and the computer expert in that they are diverted from activities that are more likely to be productive, to say nothing of the large start-up costs frequently incurred in introducing computers into a project.

During the between-session discussions, one of the problems discussed (on which there appears to be consensus) is the need for a middleman to bridge the gulf between the life scientist with no computer expertise of his own and the computer scientist who is not conversant with mathematical methodology and modeling in the life sciences. Several discussants observed that more biomathematicians are needed to fill this gap, and that this problem was not receiving the attention it deserves.

Many of the non-computer users (even those who had great difficulty understanding many of the science applications papers) expressed a very reassuring sentiment with regard to the impact of the meeting; namely, that computers can and do have a profound effect on the kind of problems addressed in life sciences research and that they need to view their own research in a way which liberates it from the pre-computer constraints of analysis and computation.

One of the important bits of wisdom that suggests a theme for future conferences is the need for individuals who understand the kinds of statistical designs and mathematical modeling so necessary to the effective application of computers in medicine. Perhaps future FASEB conferences can make a major contribution to life sciences research by focusing on the problem of integrating such individuals into the core of research involving computers in the life sciences.

Session III

Computer equipment configurations

Introduction

CHARLES E. MOLNAR, *Organizer and Chairperson*
Washington University School of Medicine
St. Louis, Missouri 63110

The plan for this session is straightforward and simple. Dr. Cox of Washington University in St. Louis will first review some trends in hardware over the past few years, and will point out some of the implications of different hardware choices on computing economy and on the kinds of problems that can sensibly be attacked. Next, Dr. Pratt, who runs the world's largest biomedical computer menagerie at the National Institutes of Health Campus in the Washington, D. C. area, will describe his operation and some of the issues involved in the choices of the 140 or so different computers that serve his large community of users.

The next three papers will describe examples of approaches to computing efficiency and convenience through development of specialized hardware. Dr. Bruce McCormick, of the University of Illinois Chicago Circle campus, will talk about a current approach to image input and analysis as applied to problems in ophthalmology. Dr. David Barry, of Washington University in St. Louis Missouri, will discuss experience in the use of macromodules as a tool for developing specialized computer systems for molecular modeling and graphics. Finally, Dr. Ted Kehl, of the University of Washington in Seattle, will describe his approach to simple and efficient computer hardware for biomedical research.

Although this session is dedicated primarily to the discussion of hardware configurations, it would be well to keep in mind the implications of various hardware choices on the kind of talent and amount of human effort needed to accomplish a research task. The experience and help available to a researcher in a given environment, and the nature of the local computing style, may well be of comparable if not dominant importance in comparison with some of the more technical issues involved in making a hardware choice.

Economy of scale and specialization—revisited[1]

J. R. COX, JR.
Washington University
St. Louis, Missouri 63110

For many years conventional wisdom stated that computers possess large economies of scale. This notion was formalized by a relation stating that a computer's power is proportional to the square of its cost (8, 10). More than a decade ago the universality of this relation (1) was questioned and succeeding events have further limited its applicability. Substantial economies have appeared in small specialized systems frequently overshadowing the economies associated with large centralized systems (5). It is hard to believe that not many years ago many computer center directors insisted that all computing within their institution be carried out on one large central machine. The success of minicomputers has forced the realization on even the computation center faithful that large systems can frequently be outperformed by smaller special purpose systems.

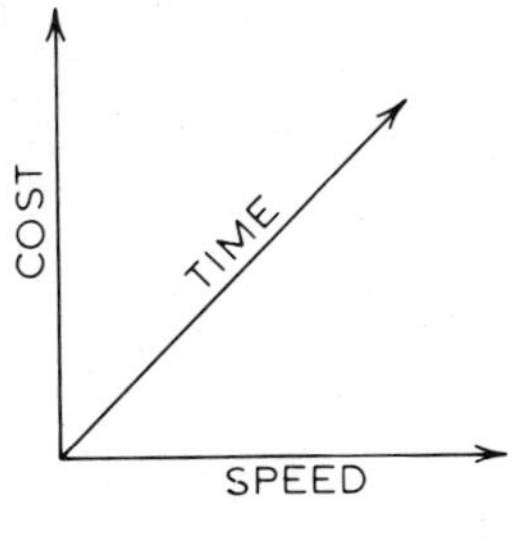

Figure 1. Three coordinates, related to the measurement of economical computing. Here cost is either the amount charged for a specific computation or a proration of the system cost to carry out that computation. Speed is measured by the number of complete executions of the computation that can be carried out per second. The third coordinate is the time required to prepare for the computation, carry out its execution, and interpret the results.

The trends that brought on this change in attitude are of interest in themselves, but moreover these trends give insight into the options available to solve today's computational problems. Economical computing can be measured by cost per computation, speed of computation, or time (or effort) required to complete the computation (Fig. 1). These three dimensions of economy have widely differing importance in different problem settings. The cost for the execution of an algorithm may appear on the bill from the computational center or may figure in the calculations carried out to justify the design of a special purpose computer system. It is the first dimension we are likely to think of in connection with economical computing. Speed, however, has importance too. Some problems cannot be approached without adequate speed. The use of computers in control and monitoring applications demands on-time performance. Other challenging and important problems may be impossible to consider be-

[1]This work was supported in part by National Institutes of Health Research Grant RR 00396 from the Division of Research Resources, Bethesda, Maryland.

cause the time to complete the computation may be months or even years.

In still other settings both the cost and speed of computation are outweighed in importance by the time to gain access to computation and the time required to interpret results. This circumstance occurs frequently for problems of modest or intermediate computational difficulty.

Let us examine the changes in these three dimensions over the past two decades. The exponential increase in speed since the mid fifties has occurred as a result of the introduction of transistor and integrated circuit technologies (Fig. 2). Each point in Fig. 2 represents a new technological advance that yields significant increase in speed. Although no

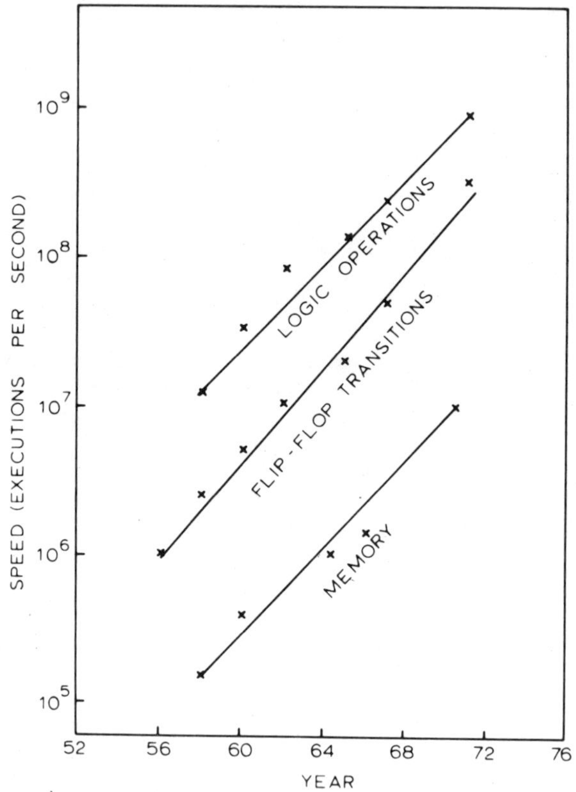

Figure 2. The speed of computer components over the past 2 decades. Dramatic increases in the speed of computation result from consistent decreases in the propagation delay through a logic gate (top curve), similar decreases in the minimum clock period of flip-flops (middle curve), and surprisingly similar decreases in memory cycle time. Vacuum tubes in 1956 gave way to transistorized discrete component circuits from 1958 through 1965. Emitter-coupled integrated circuits then took over the field of high speed logic holding sway until the present (through three successive generations MECL I, MECL II, MECL III). Memory speeds have been controlled by the size of ferrite core memory elements. This effect is shown by the left-most three points (diameters of 50 mil, 30 mil and 20 mil). Core memory speeds were pressed to their limit in about 1967 (0.5 μs cycle time) and recent speed increases depend on integrated circuit memories.

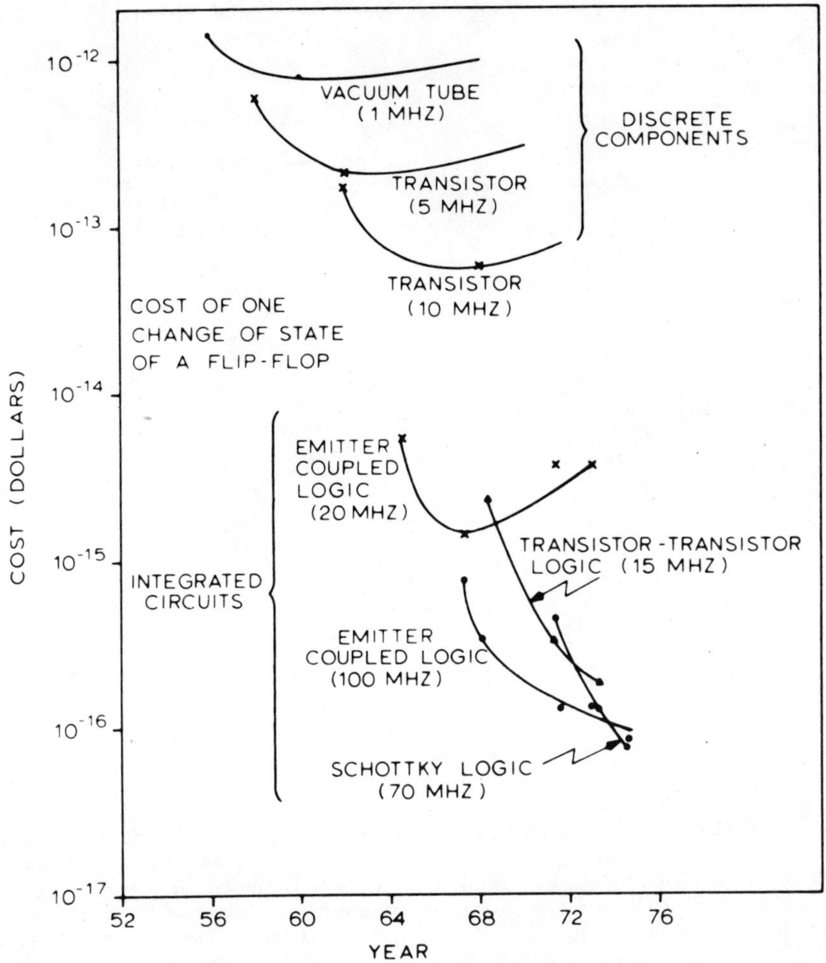

Figure 3. Cost reductions over the past 2 decades expressed as the cost of a single change of state of a flip-flop. A 40-month lifetime and continuous utilization is assumed for all the semiconductor devices. The vacuum tube curve assumes a 1-yr lifetime. Each new technology takes over from the last, increasing the volume of production for several years until high production efficiencies are achieved. A successor technology then causes the volume to decrease and, aided by inflation, costs increase. Emitter-coupled logic has been slow to reach full production despite its low cost per operation. This may be a result of unfamiliar engineering practices associated with this high-speed, current-switching line of logic.

points show any lessening of the rate of growth, no new technologies have appeared since 1971 to challenge the speed records set then.

An increase in speed coupled with a decrease in device costs has produced dramatic reductions in the cost of each digital operation (Fig. 3). The costs of Schottky and emitter-coupled logic seem to be still decreasing with their minimum still at least a year in the future. What technology, if any, will take their place

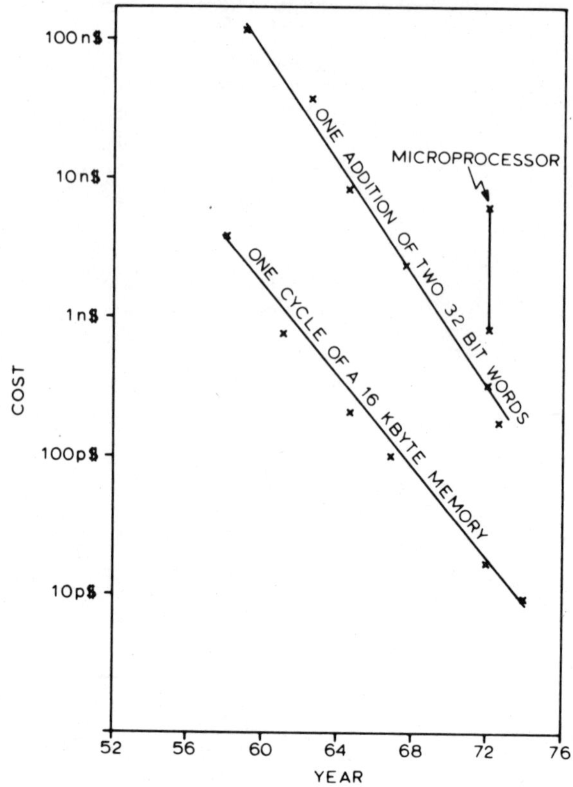

Figure 4. Cost reductions expressed in terms of the cost of a memory cycle and the cost of an addition. Memory cycles have become cheaper because of speed increases and substantial memory system cost reductions. All memory data are normalized to 16 kilobytes (kb) even though some of the earlier memories were actually much larger. From left to right the points on this curve are: TX-2 (MIT), PDP-1 (DEC), CDC 6600, PDP-9 (DEC), IBM S/7 and Fabri-Tec, Inc. All represent core memories except S/7. The cost of memory cycles as well as the 32-bit addition was computed from published monthly rentals or purchase prices (assuming a 40-month lifetime). For the addition curve continuous utilization is again assumed and the computer systems include only the CPU, an operators console and 16kb of memory. This occasionally required a rather brave extrapolation. The systems plotted are: TX-2, IBM 7094, CDC 6600, PDP-10 (DEC), IBM S/7, TI 980A (Texas Instruments). The microprocessor is an Intel MCS-8 whose cost is much smaller with the 16 kb of memory included (top point), but still is not cost effective in this comparison when no memory is included (bottom point).

remains to be seen. The decrease in cost for a single memory cycle and for a single addition (Fig. 4) covered three orders of magnitude in just under 2 decades. Pocket calculators and microprocessors because of their relatively slow speed do not yet compete with high speed minicomputers. As the level of integration increases, memory costs become a larger and larger fraction of the total cost of the computer.

The time to gain access to computation is difficult to measure. It varies from setting to setting and from individual to individual. Figure 5 shows my estimate of some of these costs over the past 2 decades. Quite obviously there is no orderly trend, a sharp contrast with the previous figures. One of the curves, however, has increased dramatically reflecting the time to read 0.1% of the operating system manuals for the Washington University system. This curve gives some indication of the dramatic increase in complexity associated with the operating systems of third generation computers.

The changes seen in Figs. 2–4 and the lack of substantial change seen in Fig. 5 have shifted the balance of concern in many settings from the cost and speed coordinates to time. Minicomputers are now often associated with new technological advances whereas a decade ago they followed the big computer leaders. To help understand the effects of these changes, I have reanalyzed the two problems first studied 7 years ago (5) but have added a time coordinate and plotted the results in a new format.

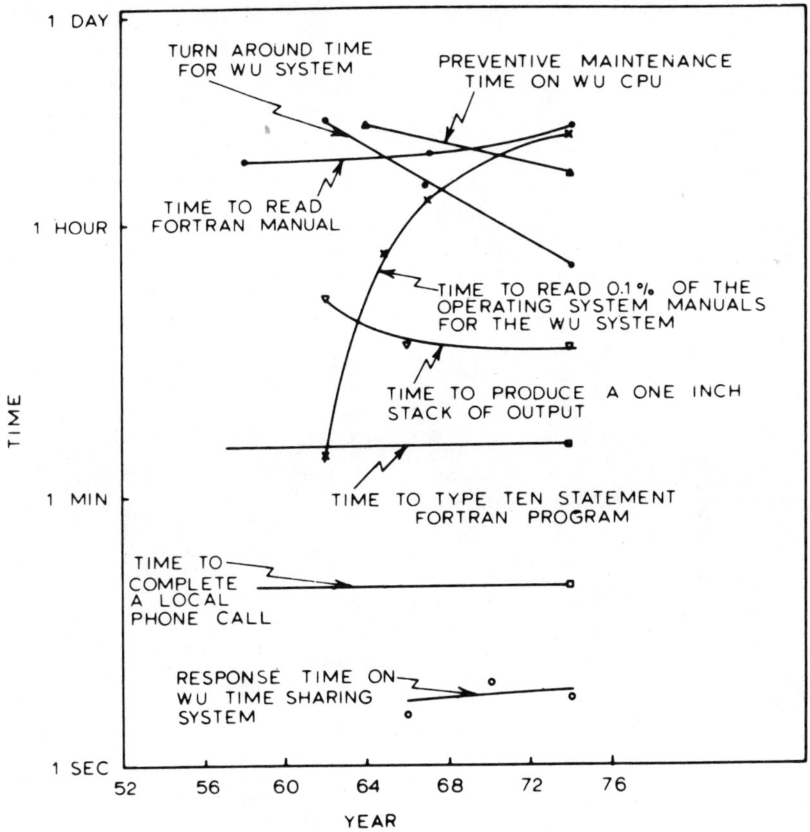

Figure 5. The time required to carry out various activities associated with computation. The data represent quite crude estimates and are shown only to contrast with the previous three figures. The activities are typical of those one might encounter in attempting to complete a computation. Many of them will take much longer than the program running time with which they are associated.

Figure 6 shows several points in the cost–speed–time space for the execution of a "butterfly" (an inner loop consisting primarily of a single complex multiplication, addition, and subtraction) of the kind used for the fast Fourier transform (4,7). The TI 980A, a minicomputer manufactured by Texas Instruments, has a system price slightly over $5,000 with 16 kilobytes (kb) of memory. The time to write and debug an assembly language program even when one is familiar with the algorithm is probably a week or a little more and is indicated by the depth from which the point is projected on the front surface of the cube. Although this is the slowest of the alternatives shown, it is considerably faster than most minicomputers and orders of magnitude faster than today's microprocessors.

Perhaps the most economical general purpose computer for the fast Fourier transform (FFT) is the CDC 6600. Its multiple arithmetic units make the calculation extremely fast and existing software makes the time to make ready short. The cost is, however, at least an order of magnitude greater than that for the TI 980A because the system cost of even a minimum configuration is two orders of magnitude greater while the speed is only a single order of magnitude greater.

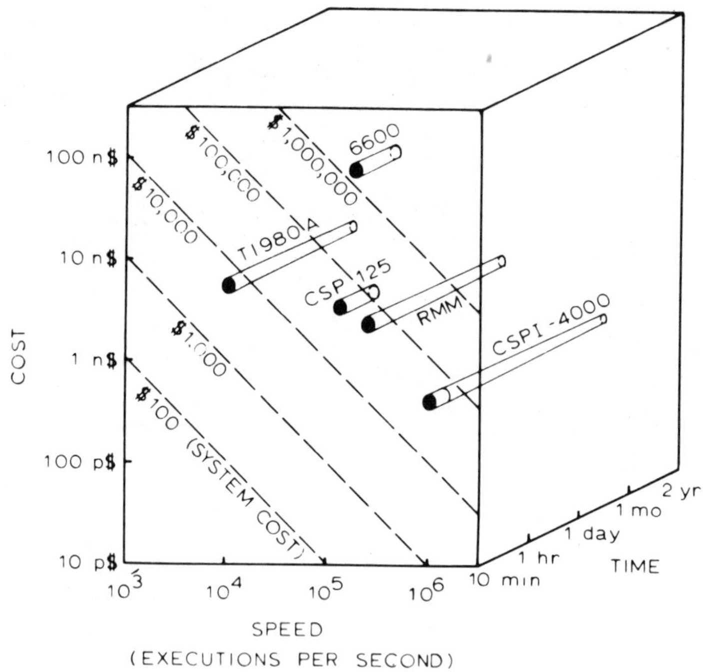

Figure 6. Three-dimensional plot of five alternative methods for computation of a fast Fourier transform (FFT). The cost is calculated for the execution of a single "butterfly," the FFT inner loop. Assumptions like those in Fig. 4 are used and the total times to make ready, compute, and interpret results are estimated from the author's experience. Total system cost relates speed to cost and this relation is indicated by the dashed 45° lines on the front face of the cube (the cost–speed plane). The data points are shown projected on this front face (black dots) from their actual locations within the cube. See the text for an explanation of the five alternatives.

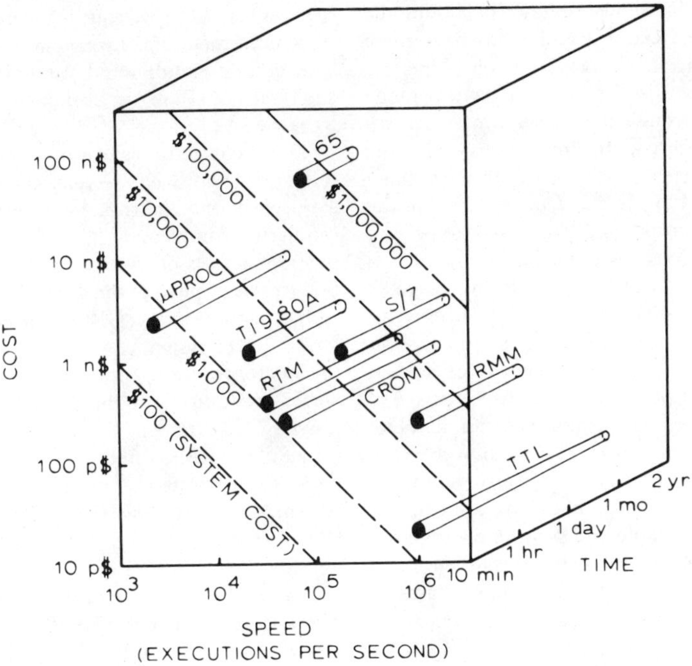

Figure 7. Three-dimensional plot of eight alternative methods for execution of the Aztec EGG preprocessing routine. The cost is calculated for a single pass through the most common path of this routine. The coordinates are as before with the data points projected on the front face of the cube (black dots). See the text for an explanation of the alternatives.

The CSP 125, manufactured by Computer Signal Processors, Incorporated (CSPI), with special features useful in FFT computation is comparable to the CDC 6600 in speed. Its system cost is substantially lower making it a clear bargain if large amounts of FFT work are required. The make ready time is probably the same or less than the CDC 6600.

Macromodules (3) in their restructured form (RMM) show an improvement in speed over the general purpose CSP 125. The cost is about the same, but it would probably take a week or two to assemble and debug a system. Of course, if a limited amount of FFT work is required, the macromodules have the advantage that they can be returned to inventory and used for other tasks.

Finally, the special FFT processor manufactured by CSPI (CSPI-4000) is both the least expensive and the highest speed alternative shown in Fig. 6. It is unlikely that many laboratories will have such a specialized processor on hand and perhaps 6 months to a year would be required to budget funds, order the instrument and install it. Once installed, its special purpose nature allows the FFT computation in minutes by merely pressing the right buttons.

Three observations regarding this format[2] for the presentation of these

[2] This format for the cost–speed plane as well as the two observations about it were suggested by C. E. Molnar.

computational comparisons should be mentioned. Most applications have some minimum speed requirement. Thus no alternatives to the left of a vertical plane passed through this minimum speed can be considered. In many situations the total system cost is a barrier to the use of a system. This is particularly true if the assumption of complete utilization is not fulfilled. We observed that a shared system (time sharing our batch) may have its system cost divided among users, an alternative that effectively moves a point horizontally to the left until the system cost corresponding to the user's share is encountered. A third observation involves the time dimension. If a maximum make ready time is specified, only points in front of a vertical plane parallel to the cost–speed plane and passing through the maximum time need be considered.

Figure 7 shows another set of points in the cost–speed–time space for the execution of the Aztec kernel (6). This electrocardiogram preprocessing program requires little numerical computation but considerable testing and branching. The slowest method shown is National's IMP-8 microprocessor (μPROC) with a system cost under $1,000. A TI 980A is again shown with only 8 kilobytes of memory at about the same cost per execution but with an order of magnitude increase in speed. A minimum configuration for the IBM 360/65 shows increased speed but is substantially more expensive for the same reasons as was the CDC 6600 of the previous figure. The IBM System/7 (S/7) because of its short memory cycle (400 nsec) outperforms all of the above alternatives, but requires a total system cost of about $20,000 with minimum memory.

DEC register transfer modules (RTM) and a specially constructed system utilizing a control read-only memory (CROM) appear close together in the cost–speed plane. The register transform module approach (2) is similar in several respects to that of macromodules, but sacrifices word length extension and speed for economy. Custom wiring is required but since each module is functionally equivalent to many individual integrated circuits, the entire job can be done in about a week.

The control read-only memory (CROM) approach decreases the complexity of the control logic otherwise required to implement a custom special purpose processor. Large scale integration can be applied to the data parts of the system so that custom CROM design may have a lower system cost (9) than any but the microprocessor approach. At least 6 weeks is required for the development of the CROM system. Restructured macromodules outperform both of these alternatives by a substantial margin but have a system cost about the same as the S/7. We have put such a system together in about a day.

Finally a custom designed and wired TTL system for performing the Aztec kernel can perform at about the same speed as restructured macromodules but with more than an order of magnitude reduction in cost. The system cost is somewhat higher than the other three custom alternatives (μPROC, RTM, and CROM). However the assembly of such a special system requires many integrated circuit packages and is likely to take more than 6 months time.

From these results one can see two and perhaps three ways to achieve economical computing. The first and most well known is to tailor the problem to existing shared facilities. If utilization of these facilities is high, economical computing results. The utilization must be high, however, over both time and resources. The more dissimilar the jobs, the more difficult it is to attain full utilization.

The second way to achieve economical computing is to tailor the equipment to the problem. Substantial economies can be achieved as we have seen, but utilization may be low nullifying any cost savings. In addition time to make ready may be undesirably high. Even so, in the right setting custom equipment has the potential for improvements in cost and speed of several orders of magnitude.

A third way to achieve economical computing combines economies of scale and specialization through the identification of computational jobs of interest to many people. Specialized equipment can be applied to such ubiquitous jobs either through high volume production or through networking many users to a large centralized resource.

The use and management of large data bases constitute an interesting and ubiquitous job that is now too expensive, in my opinion. New technology and economies of scale and specialization should bring these costs down dramatically, but whether tomorrow's system designers will accomplish this cost reduction through mass production or networking is an open question.

REFERENCES

1. Adams, C. W. Grosch's law repealed. *Datamation* 8(7): 38–39, 1962.
2. Bell, C. G., and J. Grason. The register transfer module design concept. *Computer Design* 10(5): 87–94, 1971.
3. Clark, W. A., and C. E. Molnar. Macromodular computer system. In: *Computers in Biomedical Research*, edited by R. W. Stacy and B. Waxman. New York: Academic, 1974, vol. 4.
4. Cooley, J. W., and J. W. Tukey. An algorithm for the machine calculation of complex Fourier series. *Math. Comp.* 19: 297–301, 1965.
5. Cox, Jr., J. R. Economy of scale and specialization. Presented at Future Goals of Engineering in Biology and Medicine, Washington, D. C., 1967. Reprinted in *Computer Design* 7(11): 77–80, 1968.
6. Cox, J. R., F. M. Nolle, H. A. Fozzard and G. C. Oliver. AZTEC, a preprocessing program for real-time ECG rhythm analysis. *IEEE Trans. on Bio-Med. Eng.* BME-15 (2): 128–129, 1968.
7. G-AE Subcommittee on Measurement Concepts. What is the fast Fourier transform? *IEEE Trans. A-E* AU-15: 45–55, 1967.
8. Knight, K. E. Changes in computer performance. *Datamation* 12(9): 40–54, 1966.
9. Signetics memory systems. *Economic Advantages of Microprogramming.* January, 1971.
10. Solomon, M. B., Jr. Economies of scale and the IBM System/360. *Commun. ACM* 9(6): 435–440, 1966.

NIH computer hardware complex

A. W. PRATT
National Institutes of Health
Bethesda, Maryland 20014

In recent years the conduct and management of the extramural, collaborative and intramural programs of the NIH have made increasing use of "computers." The term "computers" is used here in a generic sense since the programs are supported by a variety of computer capacities and brands. This presentation is included in the program on the presumption that a brief review of one institution's philosophy and experience in implementing computer services for its staff may be instructive to others facing a similar task. Currently, the NIH computer complement and the numbers of users of this machine complement are large; in excess of 130 machines and 10,000 users are involved. It is important to note that the effort to provide computer services to the NIH began in 1966 and involved a minimal IBM 360/40 machine and a very small number of users. The growth of this enterprise, which has been evolutionary, could be viewed as a special case in biomedical computing. Alternatively, the enterprise could be thought of as the general case, namely that a community of substantively disjoint users can, in the economic and productivity senses, profit greatly by sharing as necessary common medium and large computer capacity resources.

Currently, the NIH computer complement numbers at least 130 machines; this number does not include the machines at the National Library of Medicine which maintains an independent computer resource. Of the 130 machines, approximately 80 machines are slave input/output machines that are geographically dispersed over the contiguous 48 states with a large concentration in the Bethesda area. The machines provide remote job entry (RJE) to the NIH central computer facilities. There are two NIH central computer facilities which are designated as the IBM 360/370 and the DEC System-10 centers. The IBM 360/370 facility is a loosely coupled network of 2-370/165s, 1-360/65 and 1-370/145. The DEC System-10 is a tightly coupled dual processor PDP-10. There are slightly more than 700 registered terminals for the two central facilities. The IBM 360/370 center can accommodate 250 remote terminals on-line simultaneously and the DEC System-10 center accommodates 40 users simultaneously. The remaining machines, approximately 40 in number, are about equally divided between small capacity and medium capacity machines located in the research laboratories and clinics where each is dedicated to an explicit task or set of explicit tasks.

This simplistic "head-count" statistic of about 130 machines gains immediate perspective if one recognizes that the user cadre of the NIH facilities numbers in excess of 10,000 users who are drawn from the scientists and managers at the NIH and their counterparts in other selected government agencies; for example, the Environmental Protection Agency which has a health mission. The task or job mix spawned by this number of users, a very large number of whom have advanced scientific or managerial training, varies enormously ranging from complex mathematical and statistical computation, complex file processing, and advanced graphics to the solution of simple linear equations or correlation problems, for example, plus straightforward text-editing and report generation.

Some brief administrative, historical and philosophical comment is essential for further discussion of the NIH computer complement. The National Institutes of Health comprises what is declared to be a substantial budget partitioned across a very diverse set of programs concerned with medical research and training and biomedical communications. The magnitude and complexity of the programs cannot be easily characterized; recitation of some approximate dollar figures gives some insight, however. The FY '74 NIH budget is nearly 1.86 billions of dollars of which approximately 1.7 billions are being spent in direct support of research, training and biomedical communications. The extramural program amounts to 1.2 billions of dollars administered as several thousand grants-in-aid. The collaborative program amounts to 370 millions of dollars administered as several hundred research contracts. The intramural program amounts to 130 millions of dollars. Intramurally the NIH supports a number of basic research laboratories concerned with most aspects of modern biochemistry and physiology and also maintains a large number of clinics concerned in the research sense with all aspects of diagnosis and therapy of the diseases selected for investigative study. The magnitudes involved clearly invoke the notion that a large number of complex scientific and administrative data-handling tasks are central to the conduct and management of all aspects of the NIH program. Administratively, there is the overall need to reconcile dollar amounts with the documents and document surrogates such as research reports, individual grants and contracts, accounts payable, and the like, which define program category and procedure. Scientifically, there is an endless number of data acquisition, analysis and presentation tasks.

The first computer came to NIH in late 1958—an IBM 650. Its acquisition was justified by a consortium of administrators and epidemiologists who needed to process relatively large amounts of data rapidly and accurately according to their logical constructs. Other applications of the machine were at best visionary. The facility was operated strictly as a service. Over the next 6 years, the consortium prospered, forcing within a short time after acquisition of the 650 a capacity growth to Honeywell 800s. The processing largely represented EAM logic and the visions did not materialize. These facts led the leadership of the NIH to conclude that a strong scientific and technical base was required for developing biomedical computer applications. In order to meet this need the Division of Computer Research and Technology (DCRT) was legally established in 1964 and launched as a formal program in 1966. The mission of the administratively independent Division was to provide computer services and technical advice relating to those services and to conduct research and development programs in theoretical and applied mathematics, physical science, mathematical statistics and the information sciences. The immediate, difficult problem confronting the new Division was to prepare a plan acceptable to the NIH leadership for providing computer services and developing new biomedical applications at the NIH. The plan and the underlying philosophy that emerged, which has not, in fact, qualitatively changed since 1966, is implicitly revealed in the simplistic "head count" statistic of machines given above.

The plan called for a multiplicity of small, medium and large capacity machines to satisfy the existing and future computer needs of the NIH and related government agencies; this concept was named "distributed capacity." Implicit in this concept of distributed capacity was the notion of a hierarchical computer capability wherein the several machines could be linked together as required in a star network. Initially, it was expected that the machines would be linked together by a "shoe leather network" in which compatible magnetic tapes or punched card decks would be carried between machines.

SMALL CAPACITY EQUIPMENT

Minicomputers having as much as 8K of 12- or 16-bit words of memory have become common laboratory instruments throughout much of the world. The biomedical scientist has made innovative use of the small computer because it is free of any complex operating system, because it is relatively easily attached to transducers and other computers, and because it is relatively inexpensive. These machines have proven ideal for a variety of dedicated laboratory and clinical applications such as monitoring an experiment or a patient, controlling a complex instrument, or supervising the content and sequence of a set of basically manually executed procedures. These machines have very fast processors and are ideal for high data rate acquisition tasks. The NIH experience with these small machines is not significantly different from that of most other institutions and thus requires no further comment.

MEDIUM CAPACITY COMPUTERS

The NIH experience has shown that the medium capacity computer, the midicomputer, has proved to be a cost-effective installation for serving a small set of cooperative users who are prepared to share equipment. These machines, which may have as much as 64,000 16-bit words of memory plus a modest amount of peripheral storage and input/output gear, can be made very flexible for use by a given community of users. The machine

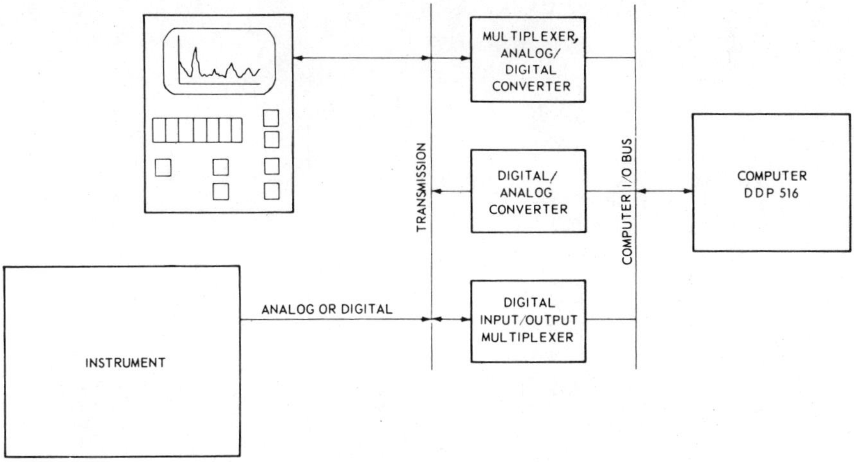

Figure 1. Typical interface of a laboratory instrument to a shared host machine.

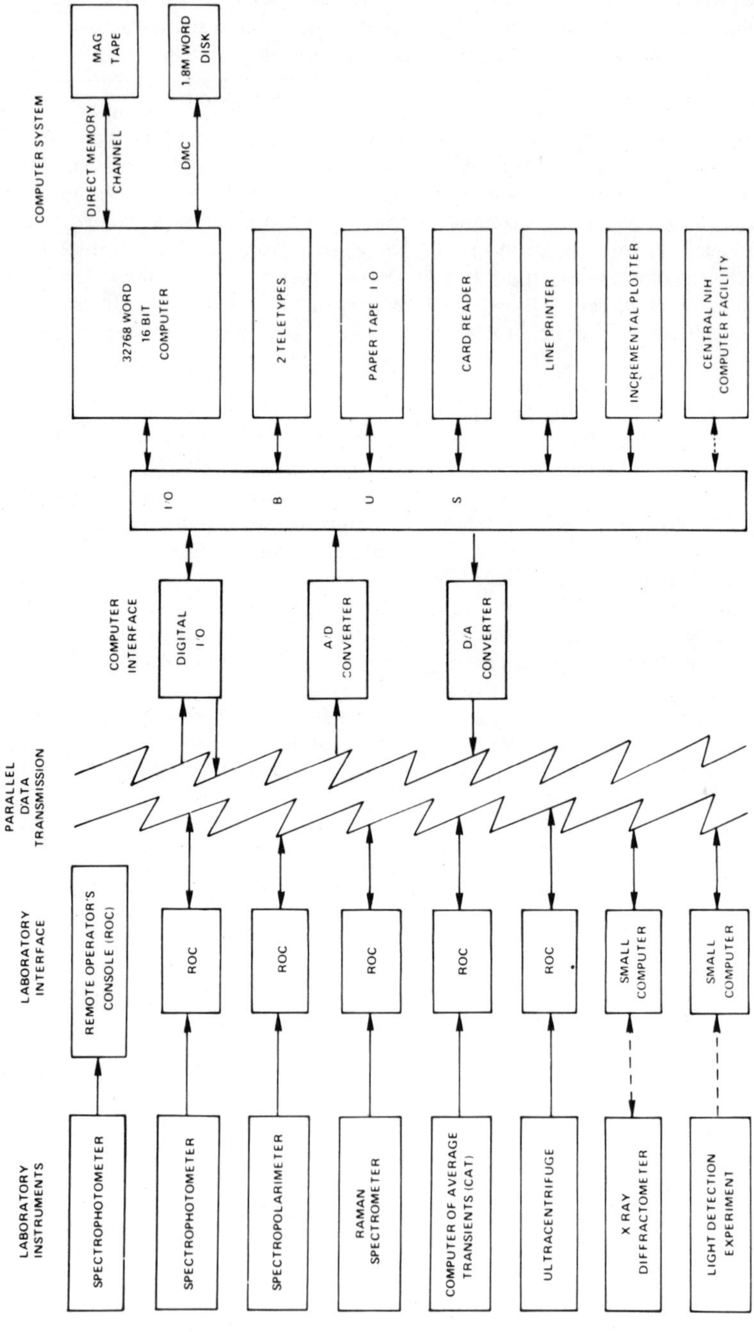

Figure 2. A schematic representation of a large, automated laboratory.

can be operated in full time-share mode and provide "simultaneously" for each of the users in a group the equivalent of a small dedicated computer for data acquisition and edit. In turn, the machine can be rescheduled to operate in the serial or batch mode and thereby provide each user rather large capacity for data analysis. Usually the medium capacity machines are operated in a time-share foreground/background mode with data acquisition and control executed in the foreground and one or more jobs such as data analysis or program compile executed in the background.

Experience has shown that almost all laboratory and clinical applications of the computer constantly evolve, requiring continuous change in the computer programs. In this aspect, the medium capacity machines have a distinct advantage. The medium capacity machine can provide the extra capacity necessary to evolve a computer application without having to disrupt existing operations. New programs can be developed in background and easily transferred to foreground. Generally the medium capacity computer that is shared among a set of tasks costs no more than the set of small machines required to handle in the dedicated mode the same set of tasks.

A typical interface of a laboratory instrument to a shared host machine is shown in Fig. 1. Since these laboratory instruments are remotely located from the host computer, a remote operator's console (ROC) is located with each instrument. These consoles are used to enter experiment parameters, to initiate, repeat or halt an experiment, and to display the edited data as acquired during a scan or an experiment. Acquired data can be stored by command at the remote console; data analysis in background cannot be initiated from the remote consoles. A schematic representation of a large, automated laboratory is presented in Fig. 2. Note that those experiments having data rates that exceed the capacity of the foreground operation of the shared host computer—X-ray diffraction and detection of evoked retinal response—are linked to the host machine by small computers. Note too, that a transmission link between the host machine and the NIH central facility is also provided (Fig. 2).

LARGE COMPUTER CAPACITY

Large computer capacity equates to some form of a centralized computer facility. In spite of the enormous amount of debate and reflection that has been focused on this subject area during the past 2 decades, no set of definitive guidelines are available to the neophite faced with the task of establishing a central computer facility. Thus, defining and implementing a central computer facility is a difficult and thankless task. The task is difficult because there is no relatively inexpensive, universal computer. On the contrary, the almost mutually exclusive nature of interactive and batch processing suggests the antithesis of universality. One must always seek some balance between these two forms of computing and almost always over the vehement protests of the involved protagonists of the different forms of computing. Also, and contrary to prevailing opinion found among government, campus and corporate administrators, there is the fact that no vendor's software and hardware satisfies the custom needs of an organization. One must always innovate with the vendor's inventions to achieve good performance. Recounting the NIH experience in establishing central computer facilities is done only to add another small increment to history in hopes that history in some way aids in contemporary decisions. Thus, it is of interest to list some of the facts and perceptions that in the aggregate determine the philosophy that undergirds the NIH central computer facilities.

1) A well organized batch stream is necessary to accommodate compute-bound tasks in mathematical analysis and complex file processing.

2) In addition to traditional modalities, interactive subsystems for job definition (data definition, programming, and so on) and easy job entry into the batch stream are mandatory.

3) Job segregation by programming language—COBOL versus FORTRAN—and/or by substance—administrative versus scientific—is wasteful of central facilities, since load balancing is difficult and there is often unnecessary duplication of staff and facilities. The NIH scientists and administrators share the central facilities since they are convinced that neither group could fund separately the needed facilities for each group.

4) The central facilities must operate on a cost-recovery basis, that is, costs of equipment, personnel and supplies would be recovered through fees charged for services rendered.

5) The central computer facilities must provide a powerful graphics technology. The reason for this requirement is straightfoward. Biomedicine is not a theory-driven science and thus cannot to any large extent be substantively represented by analytical mathematical expressions, e.g., $y = (f)x$. The biomedical scientist is forced to recognize and in turn represent data by graphic displays such as lists, plots, or microscope images. For this reason, computer manipulation of biomedical data needs to be augmented by graphics.

As noted earlier, the NIH operates two central computer facilities. The larger center—IBM 360/370 central computer facility—represents a successful experiment in designing and implementing a computer utility in a biomedical environment. The heart of this general purpose utility is a large IBM 360/370 multiprocessor system, operating in a multiprogramming mode and providing batch processing plus time-share and graphics services. The utility operates on a fee-for-service/cost recovery basis on a 24-hour daily schedule. Approximately 5,000 jobs—95% of which are submitted by

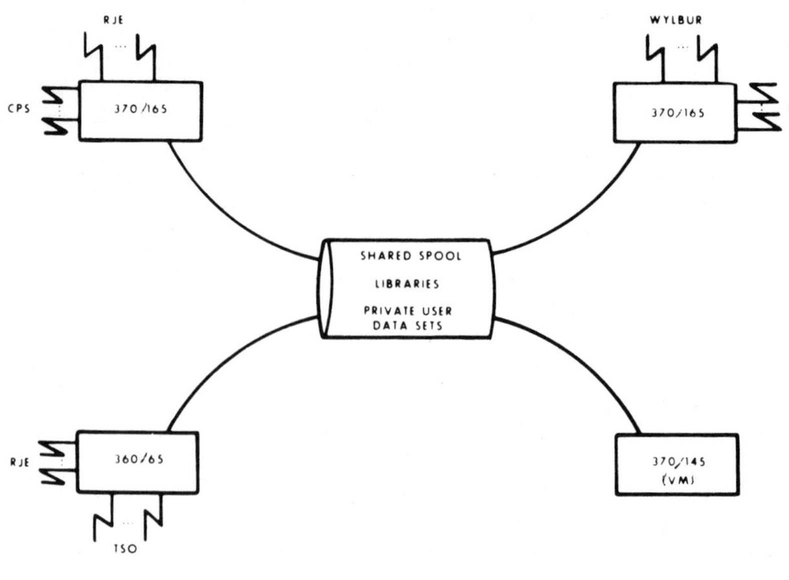

Figure 3. Shared spool—NIH configuration, May 1973.

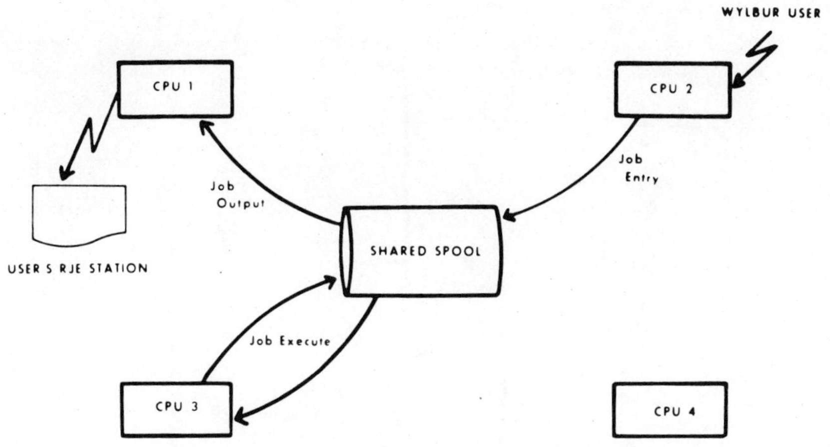

Figure 4. Shared spool—job path.

remote job entry or from remote keyboard consoles—are processed daily.

The utility, which consists of 2-370/165s, a 360/65 and a 370/145, is arranged as a loosely coupled network in which each participating processor functions independently of the others, accepting jobs for processing solely as a function of its available capacity and access to required devices (Fig. 3). Each machine can perform all stages of job processing and different stages of a job can occur on different machines (Fig. 4). The software which creates the loosely coupled network is known as the NIH Shared Spool and is a major modification of the IBM HASP 3.1. Under this software, up to seven processors can share HASP's spooled disks. All communications are via shared direct-access-storage devices; no channel-to-channel adapters are employed. Since no master/slave relationship exists, the failure of a machine or putting a machine off-line does not crash the entire network. Load balancing and thus increased throughput is automatically accomplished.

The NIH WYLBUR system which is an interactive text editing remote job entry system is an essential component of the utility. The original WYLBUR kernel was written at Stanford University (R. Fajman and J. Borgelt, Comm. ACM V.16, N5, May 1973) and has been extensively expanded at the NIH. WYLBUR is a comprehensive system for manipulating all kinds of text such as computer programs, letters, manuscripts, using keyboard terminals connected to a computer. In addition to the facilities for text alignment and justification, the system provides facilities for retrieval and addressing text by content. The system also serves as an interactive, nearly transparent interface to the batch stream. Slightly more than 200 users can be on-line simultaneously.

The DEC System-10 computer facility is a much smaller facility and centers on a fully time-shared, dual processor PDP-10. The facility is strongly oriented toward computer-based graphics. The system which is schematically represented in Fig. 5 supports approximately 450 scientific users who are concerned with difficult pattern recognition problems and mathematical modeling. Up to 40 users can be on-line simultaneously. In addi-

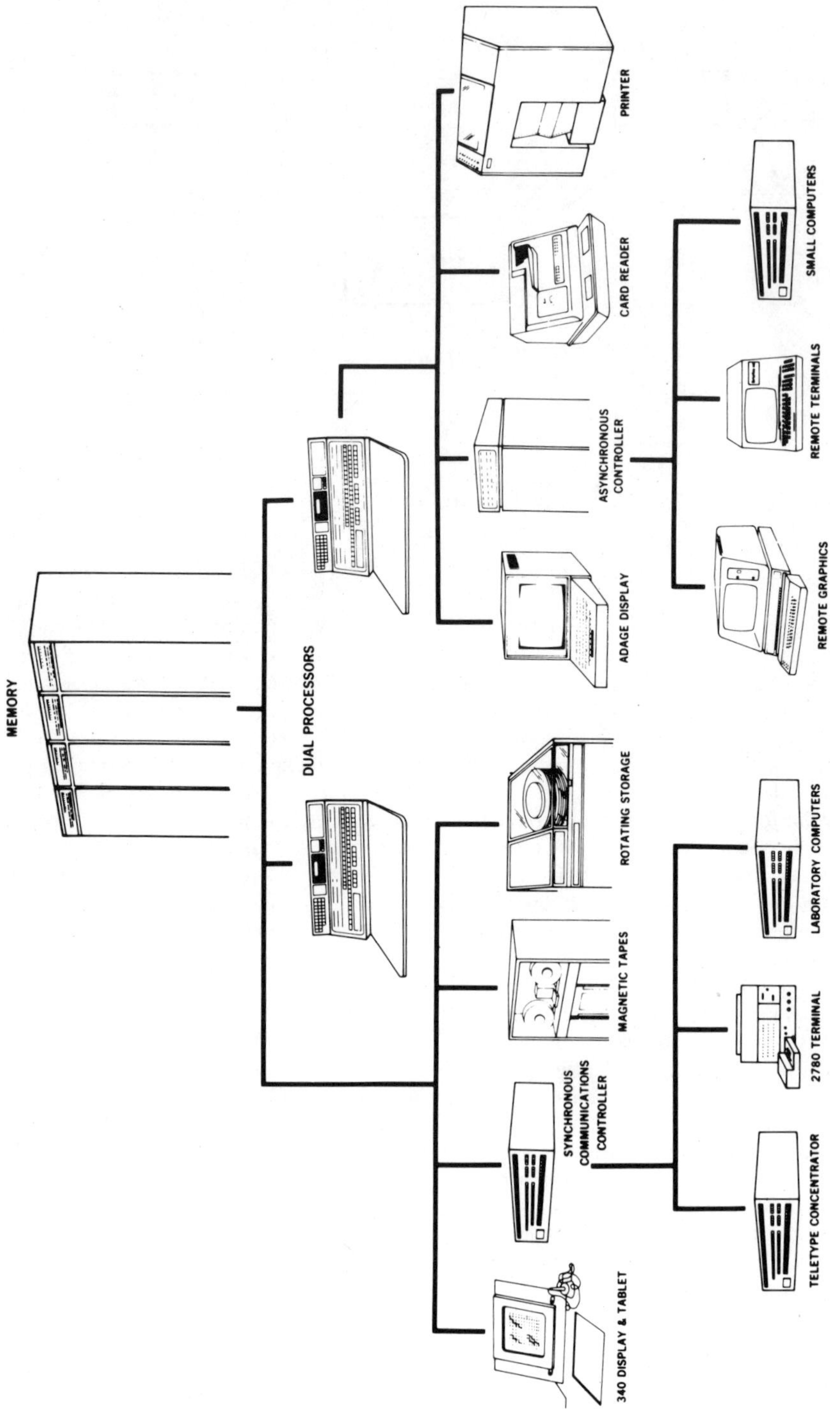

Figure 5. DEC system-10 computer facility.

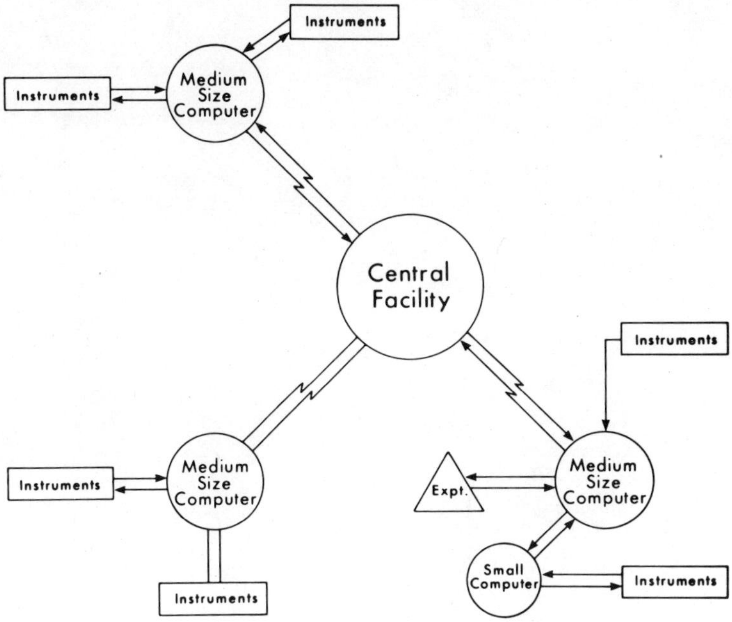

Figure 6. Network—peripherals supported by central facility.

tion, a number of minicomputers communicate directly with DEC System-10 in a true form of hierarchical computing (Fig. 5). Two major NIH software packages greatly enrich the system. The first is OMNIGRAPH (written by R. Sproull) which provides broad graphics support to the users. Display and/or plotting of data is made independent of the programming language and graphics device used. The second software package is MLAB (written by G. Knott) which supports much of the mathematical modeling work at NIH. MLAB is a computer language that greatly assists a user in curve-fitting of data and solving differential equations.

Some effort has been made to create a star network among the small, medium and large capacity machines (Fig. 6). In fact, software for transmitting "buffers" of data between three medium capacity machines and the large central computer facility exists. However, the network is currently inflexible and inefficient and warrants no more than passing mention.

Hopefully two points that should be of interest to the biomedical community have been developed by this nontechnical description of the NIH computer hardware complement. *1*) The conduct and management of biomedical research requires a variety of computer capacities and configurations, and *2*) If the biomedical managers and scientists will modify their devotion to autonomy and pool their limited financial resources, it is possible to have available, almost on demand, powerful computing resources.

Image processor for biomedical research[1]

B. H. McCORMICK, R. T. BOROVEC
J. S. READ AND R. C. AMENDOLA
University of Illinois at Chicago Circle
Chicago, Illinois 60680

1. THE TELEVISION OPHTHALMOSCOPE APPLICATION

The experimental computer-driven Television Ophthalmoscope system (TVO) is an important new source of quantitative information about the eye. Data of considerable potential clinical importance are obtained via computer analysis of television images of the ocular fundus (back of the eye). This method is noninvasive and causes minimal discomfort to the patient. A TV camera views the fundus through the pupil of the eye via a standard clinical fundus camera. A special purpose digital image acquisition and processing system performs basic data reduction tasks as well as those data analysis tasks requiring interaction with an operator.

The proposed clinical use of the system is in *1*) time-flow studies of the retinal circulation, *2*) spectral analysis of features in the fundus image, and *3*) geometric analysis of the fundus image, e.g., the optic disc and retinal vessel configuration.

All three TVO measurement capability classes (time-flow, spectral and geometric) will add precision and accuracy to diagnosis and consequent decision-making in the retina clinic. For example, repeated time-flow studies in diseases such as retinal venous occlusion would offer a means of determining whether circulation to the affected area is improving or becoming worse. This objective information could be of predictive value as to when venous occlusion should be treated, i.e., by drastic therapy such as laser photocoagulation. A broad range of clinical studies has been planned to evaluate the usefulness of TVO-derived image data.

2. DESCRIPTION OF IMAGE PROCESSOR

The image processor comprises three parts: the Minicomputer, the Image Memory, and the Video Controller (see Fig. 1). Most of the system is constructed from relatively inexpensive, commercially-available assemblies.

Minicomputer

The computer consists of a general purpose minicomputer and peripherals. The Lockheed SUE was selected for several reasons:

1) It has a high-bandwidth (5 million words per second), general bus architecture. Bus arbitration is performed by a bus controller, and configuring a multiprocessor system is simply a matter of plugging in another processor.

2) It has a microprogrammable processor control.

3) It is inexpensive.

---

[1]This study is supported by NIH Research Grant # US PHS EY 01261-01.

The SUE is used as a dual processor machine. One processor is an unmodified scientific processor that features extensive bit manipulation instructions as well as fixed point arithmetic operations. The second processor is a general-purpose processor modified to use a writeable control store for execution of custom instructions. We will use this feature to increase the execution speed of critical image-processing and graphics functions.

Peripherals in the system include, in addition to the usual disk, tape, and alphanumeric CRT console:

- A pen-type digitizer for rough digitization of images and as the interactive medium for graphics.
- A high speed printer/plotter for listings and halftone representations of images and graphics.
- A removable-cartridge video disk.

Image memory

We believe very strongly that the most reliable method of handling images is in the digital form. A fast, large core memory is used to provide temporary and working storage for image processing. This Image Memory is also used to refresh images on the displays. With the advent of second sources for mass memories for large-scale computers, and the very competitive market for minicomputer memories, this decision to use core can be economically justified. The principal advantages of core over disk include: higher speed, and greatly improved flexibility in accessing the picture. In addition, the video system does not have to be synchronized with the core memory as it does with a disk.

The image memory (see Fig. 2) comprises eight memory modules (initially 8,192 words by 16 bits per word) which are accessed in parallel to give an effective 128-bit word with an approximate cycle time of 750 nanosec (for a bandwidth of 171 megabits/sec). The same address is accessed in each module.

Since it is necessary for more than one

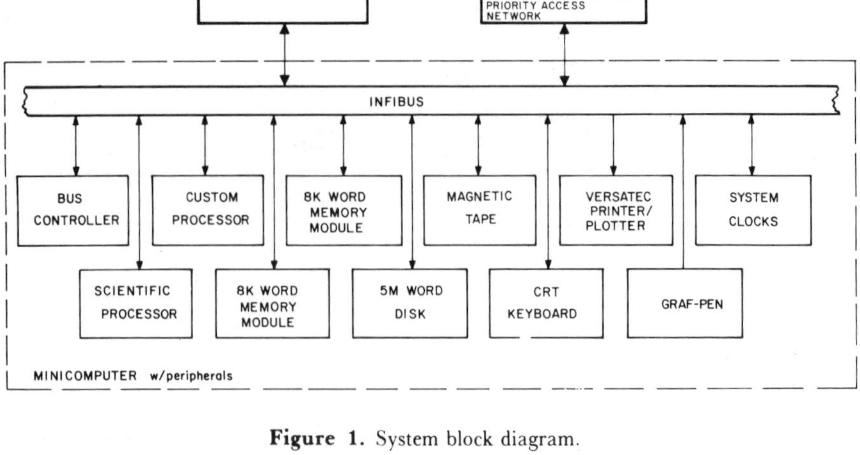

Figure 1. System block diagram.

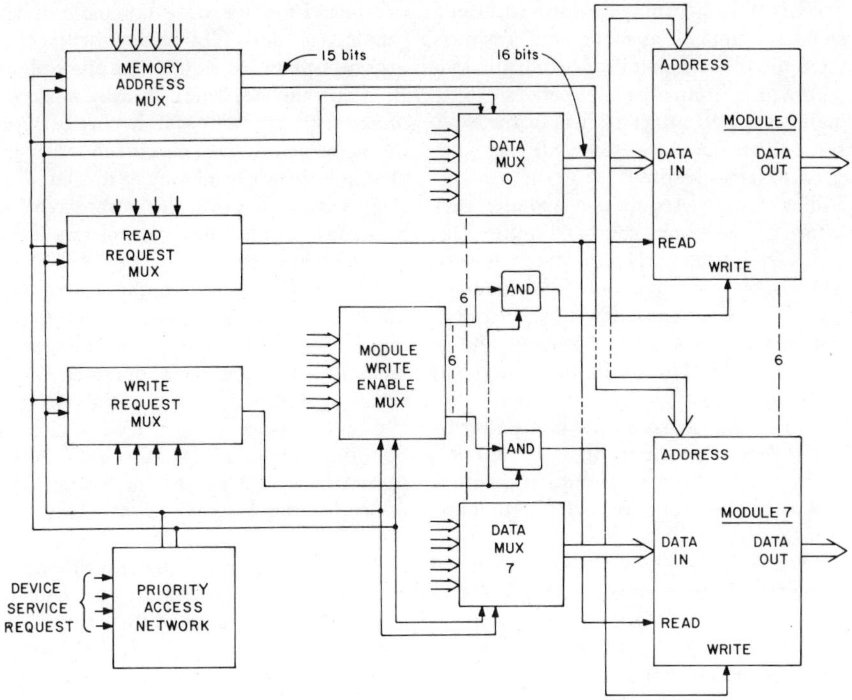

Figure 2. Image memory schematic.

device (e.g., the video controller, the minicomputer, etc.) to access the memory, a four-port Device Input Multiplexor (DIM) is provided. Access to the DIM is provided by the Priority Access Network (PAN). A device requests access by asserting its Device Request Service (DRS) line and a three-bit Device Priority Number (DPN). Any device asserting a DPN higher than the active DPN will terminate the active device access. Two DPNs are permanently assigned; the video controller in the digitize mode has the highest priority, the video controller in the display mode has the lowest priority.

The Bidirectional Scratchpad Array (BSA) serves as the memory data register for the image memory. It comprises a 128 bit register that may be accessed as either eight 16-bit words (horizontal mode) or sixteen 8-bit words (vertical mode) (see Fig. 3). Cycle time for an access operation is approximately 250 nanosec. For communicating with the minicomputer the BSA is treated as an extension of the SUE memory address space. The bidirectional access capabilities are useful in certain classes of image processing since one has available both a "planar slice" of an image (horizontal mode) or the gray scale value for a given pixel (vertical mode). Each word in the horizontal mode corresponds to a module in the image memory (see Fig. 4).

Video controller

The Video Controller accepts a video (analog) input and stores its digital representation in the image memory. It also accepts digital image representations from the image memory and converts them to video for display or transmission. The video controller consists of three main parts, all under control of the SUE system: Image Acquisition, Image Display, and the Digitizer.

The Image Acquisition module performs all necessary signal conditioning, e.g., separation and normalization of video and sync portions of an input composite TV signal. The image acquisition module is device dependent and is custom designed for each different type of video source.

The Image Display module performs the inverse operation, that is, it takes normalized video signals and sync signals and transforms them into device compatible signals.

Quantized graphics are handled in the system as if they were 1-bit gray scale images. The system allows the interpretation of two arbitrary planes (image memory modules) as graphics.

A video mixer for each display allows the arbitrary mixing (under program control) of the following signals: *a*) video (analog) or video (digitized); *b*) cursor; *c*) graphics plane A; and *d*) graphics plane B. Thus the user may overlay a stored image with graphics and display both on the same monitor. The cursor is generated in hardware and may be displayed as a set of crosshairs over the entire screen or may show as fiducials at the corners of the windows of interest.

The Digitizer converts the analog output of the image acquisition module into digital form for storage in the image memory. The digitizer operates on a raster of 1,024 by 1,024 points with a one-to-one aspect ratio. Devices such as 525 line television systems (with nominal resolution of 480 by 640 lines) fit into the upper left hand corner of the raster (see Fig. 5).

Two modes of operation of the digitizer are provided. The first allows selection of one arbitrary window in the raster. The user specifies the begin and end coordinates in the raster as well as the resolution (distance between sample points) in X and Y. The second mode of operation

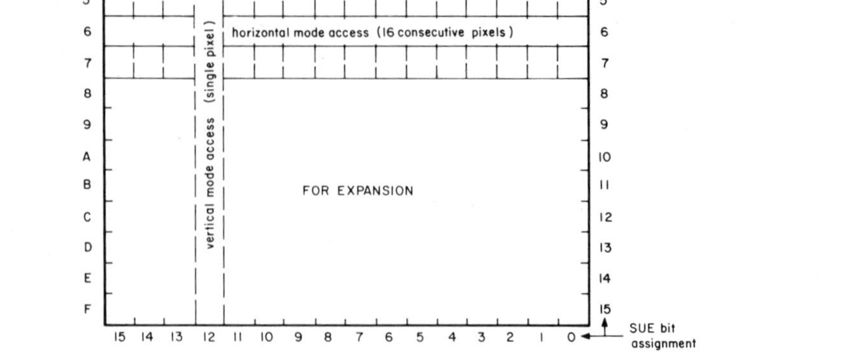

Figure 3. Bidirectional Scratchpad Array (BSA): addressing conventions.

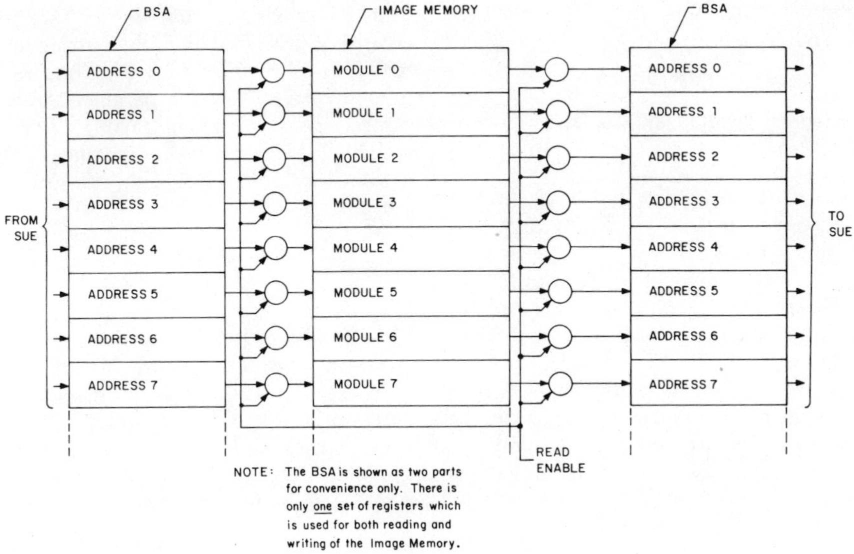

Figure 4. Image memory module versus BSA horizontal address.

allows the user to specify up to eight small subwindows in one frame. This allows tracking of several areas at once. Due to the speed requirements of the digitizer there are some restrictions put on the placement of these subwindows in the raster. In the multi-window mode the subwindows are specified in a display list whose format resembles that of the single window parameters. That is, the coordinates of the begin and end points are specified for each subwindow. The resolution specification holds for the entire frame.

Memory management is handled by the digitizer hardware. The user specifies the beginning image memory address at which he wishes the image to be stored. For multi-window mode, as many as 8 separate beginning addresses may be specified. The address is then incremented at the appropriate time by the hardware as it accumulates 16 pixels at the desired gray scale. The user may specify 1, 2, 4, or 6 bits of gray scale for any frame. He also then specifies which sets of memory modules are used to contain that gray scale information.

The conversion of digital images to analog images is handled in the same manner. The user specifies the window and the initial image memory address and that image is accessed and converted. The user may also manipulate the placement of images by changing the begin and end points appropriately. This allows the mosaic construction of small images into one large image. For example, it is possible to simultaneously display an original image, images generated at intermediate processing steps, and the final result.

3. THE SYSTEM LANGUAGE–APL/EYE

We are implementing a dialect of APL (APL/EYE) as the principal system language of the TVO system. APL/EYE contains a considerable part of the IBM APL/360 implementation as well as several extensions specific to the TVO system. We plan to use APL/EYE to develop tailored application packages to perform specific ophthalmological research tasks.

Image acquisition and display

The video controller hardware features described in *section 2* are supported under APL/EYE.

To meet the performance requirements of video angiography and other applications involving scan/display of dynamic imagery, APL/EYE includes a movie facility. The movie facility controls rapid movement of sequences of pictures between a picture source or display and either the analog or video disk. Individual movie frames can be stored or retrieved from either disk so that movies can be created or processed digitally. During the making of a movie with input from the fundus camera, several parameters are controlled by the movie facility. These include TV camera beam blanking, light dose energy and wavelength, and frame scheduling.

Image processing

Most image processing operations have been implemented within the existing APL syntax. APL provides a concise notation for specifying operations on arrays. For example, one can add corresponding elements of two arrays with a single APL statement. APL's arithmetic

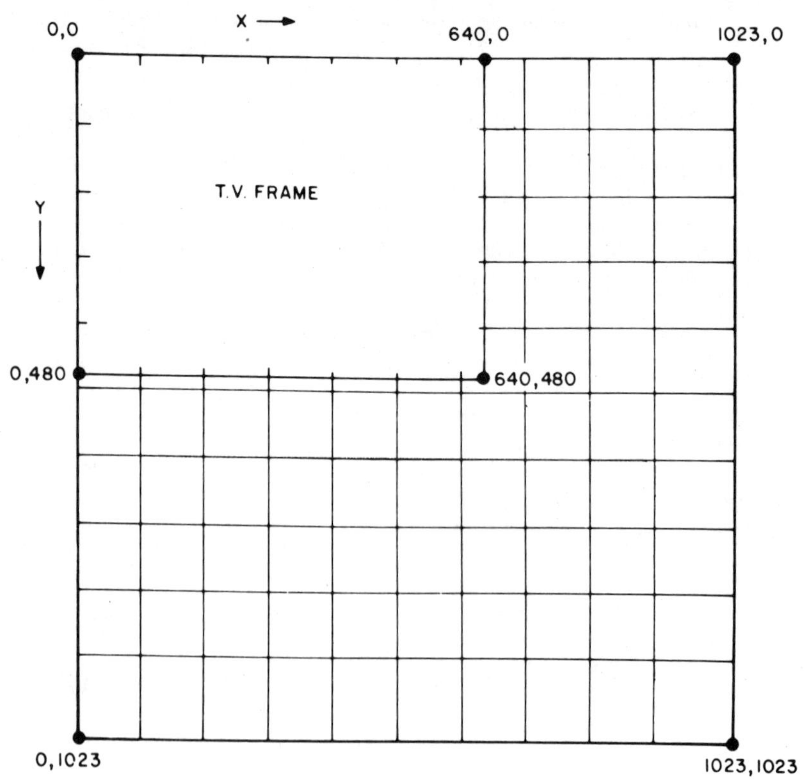

Figure 5. Digitizer raster field.

and logical operations may be used to process pictures that appear to the user to be large two-dimensional integer arrays of varying precision. Two additional operations, *convolve* and *count* have been added. *Convolve* performs the convolution operation between two pictures, thereby effecting a large number of linear position-invariant picture operations such as shifting, smoothing, noise-cleaning and template matching. The *count* operation applied to a picture returns the number of picture elements having a specified gray value.

Graphics

APL/EYE includes basic, though complete, facilities for generating two-dimensional computer graphics. The usual capabilities for point, line, or text plotting, scaling, and windowing are provided. All graphical output is buffered in the image memory, and can be displayed on the TV monitor or printed on the printer/plotter to produce hard copy. Facilities are also provided for use of the GRAFpen digitizer in two modes: as a data input device, and as a means of controlling program execution. APL/EYE can quickly switch back and forth between these modes in a facile interaction with the operator.

4. DISCUSSION

Our experience has been that minicomputer-based image processing systems have been shown to be highly cost-effective and can compete very effectively against conventional large scale computer systems. The system presented here can be given the following thumbnail characterization: it provides image acquisition from 3–5 conventional television frames per second for a total sequence length of typically 20–30 pictures; television images are then encoded in 64, 16, or fewer gray levels; subsequent image processing can be pursued at leisure but normally must be completed in 10–20 min. Digital image processing is used throughout—to ensure a stable and reliable instrument capable of sustained operation in a clinical environment.

The system described above has been tailored for specific uses in ophthalmoscopy. The input devices are standard television cameras while the output devices are standard television monitors. There is no reason, however, that the acquisition module could not be changed to normalize the output of other imaging devices. The rest of the hardware, as well as the APL/EYE system software, is general and transferable to other applications. Analogous biomedical image processing applications which have in common a need for a low-cost, modular image processing system include: *1*) examination of lung radiographs, including node detection and pneumoconiosis screening; *2*) automated cytology, including karyotyping of chromosomes, differential counts of white blood cells, and characterization of Pap smears; *3*) xeromammography for detection of breast cancer; and *4*) brain mapping—neuron reconstruction from electron micrographs, neuron morphology as identified by heavy metal, fluorescent or conventional stains under light microscopy, and gross anatomical reconstruction of the brain.

Evolving macromodular molecular modeling system[1]

C. D. BARRY, H. E. BOSSHARD
R. A. ELLIS AND G. R. MARSHALL
Washington University Medical School
St. Louis, Missouri 63110

This paper deals with the role digital computer building blocks, such as the "macromodules" developed at Washington University, St. Louis,[2] can play in providing the scientist with flexible, high performance computing facilities. The development and use of a system for molecular modeling has been chosen for illustration.

The computer model essentially consists of atomic coordinates that can be transformed by operators representing the effect of bond rotations while preserving the stereochemistry of the molecule. The model is normally used to identify molecular structures consistent with experimental evidence, the fitting of the model to observed data being pursued with a computer either interactively, following the "intuitive, trial and error" style used with mechanical models, or automatically by defining some criterion function to describe the acceptability of a conformation which can then be optimized (15, 21, 22). In the interactive mode, which is able to take advantage of the powers and versatility of the "human computer" for pattern recognition and inductive thinking, the model is usually presented as a three-dimensional line drawing using a computer controlled display (2, 13, 14, 17, 19). The rationale is to produce an illusion (a hand–eye correlation) strong enough that the operations required to manipulate the model via the computer become instinctive. The success of the illusion is closely related to the smoothness of the interaction rather than to the sophistication of the molecular portrayal[3]. As the smoothness can be related directly to machine performance it serves to establish the computational demands for effective real time interaction.

Four major computational tasks can be identified.

A) Display. This involves generation of lines and control of the line drawing scope. In most systems these functions are carried out by a specialized computing element—a Display Processor.

B) Computation associated with static displays. This involves the conversion of the model data into the form required for display and also any overhead associated with control of the display processor.

---

[1] This project was supported by a National Institutes of Health grant RR 00396 and by contract SD 302 from the Advanced Projects agency of the Department of Defense. One of the authors (GRM) is also supported by an Established Investigators award from the American Heart Association.

[2] Macromodules have been defined as "building blocks such as registers, adders, memories, control devices, etc., from which it is possible for the electronically-naive to construct arbitrarily large and complex computers that work" (7).

[3] The quality of the display must be high but the content can be simplified.

C) Computation associated with dynamic display[4]. This involves provision for scaling, translation and rotation (reorientation) of the display as a whole on a frame-to-frame basis.
D) Manipulation of the model[4]. In the context of molecular modeling this involves the generation of different conformations of the molecule for display. It should also include calculation of the new atomic coordinates that can be used to evaluate the conformations.

Both categories *C* and *D* require matrix operations that can rotate the molecule (or display) as a whole, or rotate selected atoms about a bond-axis to generate a new conformation. In an application where these tasks represent a major portion of the computation, provision of special facilities for matrix multiplication with have a significant effect on performance[5]. Similar, though less spectacular savings can be obtained by using specialized hardware, when two or more portions of a calculation can be found that are independent, allowing parallel processing, where specialized memory organization or addressing can be tailored to the data structure to optimize access, or when registers can be provided to hold frequently used data. Provision of suitable program elements as hardware routines brings, as an additional effect, a saving in memory references as the program itself resides in hard-wired control sequences rather than as program codes that must be read from memory.

MMS-4 represents the present stage in development of our system for use primarily in the study of macromolecular structures (proteins). It is used actively in the accurate presentation and interpretation of electron density maps (12) obtained by X-ray crystallography (Fig. 1), in the display (Fig. 2) and detailed examination of known protein structures (16)

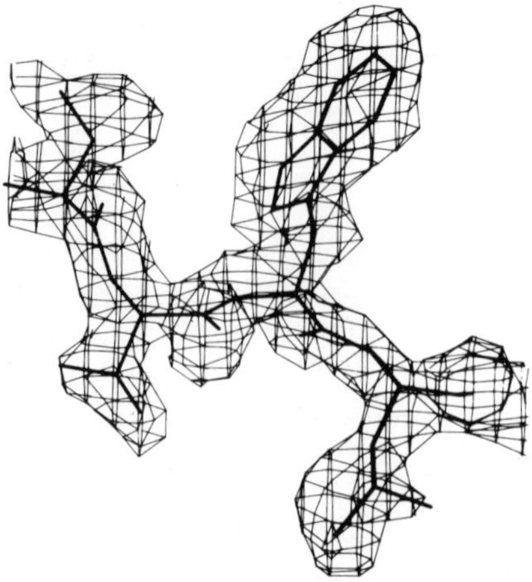

Figure 1. Fitted electron density map of the Try 22 region in cytochrome b_5. (Courtesy F. Scott Mathews).

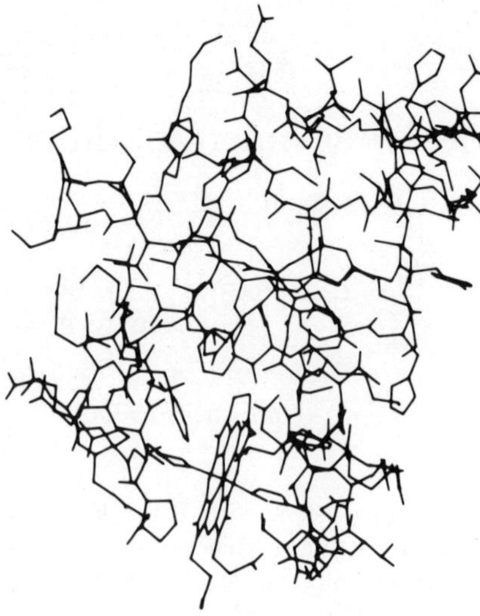

Figure 2. The display of cytochrome b_5 showing the heme group and all non-hydrogen atoms. This represents about 25% of the MMS-4 capability.

[4] These include the monitoring of input devices related to these operations.

[5] This is often implemented as part of the display processor, instructions relating to matrix operations on portions of the display (subpicture elements) being embedded in the display list.

Figure 3. MMS-4. The components from left to right are *a*) auxiliary memories, *b*) macromodule processor, *c*) E and S matrix multiplier, *d*) human computer, *e*) Linc scope and keyboard, *f*) E and S display and *g*) Micro Linc-300.

including the production of cine films for educational purposes, and in general model building using existing structures to postulate the conformation of proteins with homologous sequences.

The system architecture is unique, drawing heavily on the use of hardware routines implemented with macromodules (10). Three major subsystems can be identified (Figs. 3 and 4).

1) A small, general purpose stored program computer, a Spear Micro Linc-300, with 8,192 words of 12 bit, 1 μsec memory, a small point plotting scope, a keyboard, potentiometers for analog input, a teletype and four small magnetic tape units (Linc-Tape) for mass storage of programs and molecule data.

2) A portion of the Evans and Sutherland Corp. (E and S)LDS-1 system comprising the high speed line generator and scope which incorporates display windowing and depth cuing (intensity modulation depending on display depth) and the digital 4 × 4 matrix multiplier capable of producing a transformed 24-bit, four component vector every 8 μsec.

3) A macromodular (MM) machine assembled from 139, 12-bit control modules incorporating 73,728 word of 2 μsec, 12-bit memory and with interfacing to the Linc and the Evans and Sutherland hardware.

Overall control of the system is provided by the Linc, (the only element of the system that can be programmed in the conventional sense) through a set of options presented as menus on the point-plotting scope that comprises part of the Linc (Fig. 5). Selections are made using the Linc keyboard. As these functions only require response times short on the human time scale and also represent portions of the system most likely to change as the system evolves, the use of the software here is an obvious choice.

Frame-by-frame interaction with the display or model through the parameters associated with various potentiometers is more demanding on the system, requiring response times of the order 50 msec and the flexibility associated with software. Experience with another system in England (2) has shown that even with a moderately fast machine and independent display facilities the software approach is limited to structures with less than 200–300 atoms (Table 1 and Fig. 6).

The MMS-4 solution has been to provide the special purpose hardware routines of the type mentioned earlier, with the Linc providing their control. The commercial production of the high performance E and S hardware removed the need to develop functionally similar elements and favored its inclusion. Macromodule routines provide direct control of the multiplier, commands being generated by the modules together with the fixed, fourth component of the input vector (20). The coordinate data flow between the multiplier and the external memory involves separate portions for input and output which are handled by two asynchronous hardware loops allowing maximum performance.

Both the display of the molecule and the application to the appropriate atomic coordinates of a single transformation matrix corresponding to rotation about a bond in the molecule, have been implemented as self-contained hardware se-

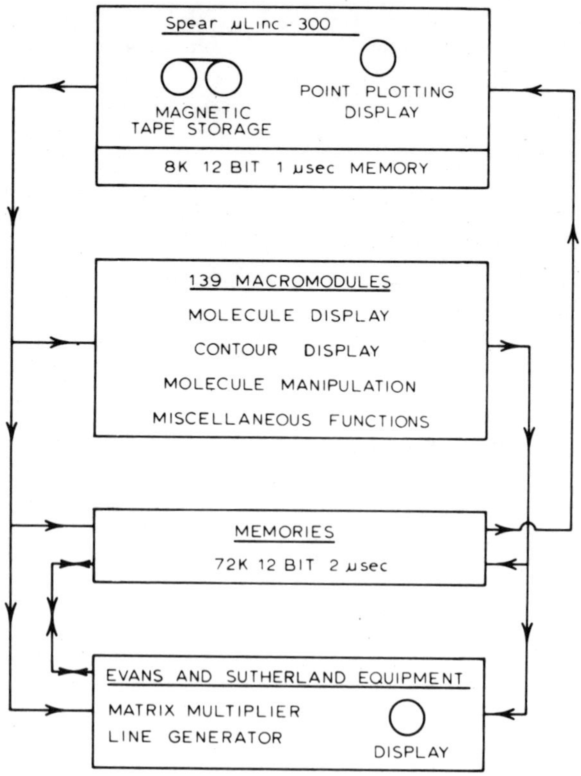

Figure 4. Block diagram of MMS-4.

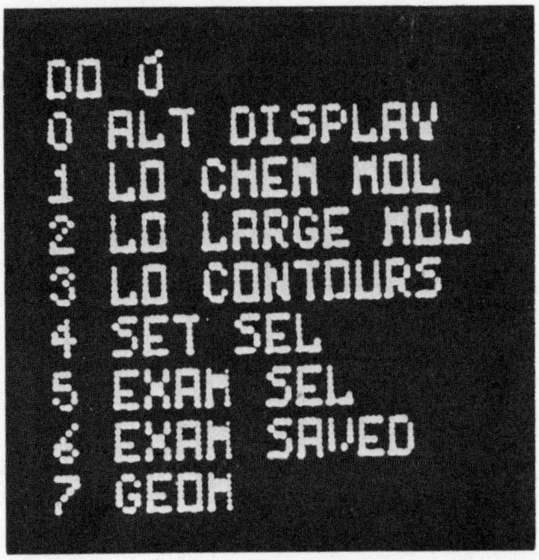

Figure 5. Example of menu control displayed on the Linc scope.

quences requiring only the transfer of the appropriate matrix and a pointer from the Linc before initiation. The Linc is then free for calculations until the hardware sequence is complete (Fig. 6). Data such as atomic coordinates and connectivity, atom types and data specifying the atoms transformed by bond rotations, required by these hardware routines, are held in the 2 μsec, marcomodular memories. These are external to the Linc and so allow each machine independent access to the appropriate memory without the need for cycle stealing. Transfer of data between these memories occurs primarily when a new molecule is entered from the data stored on Linc tape. The transfer is handled as programmed input/output,

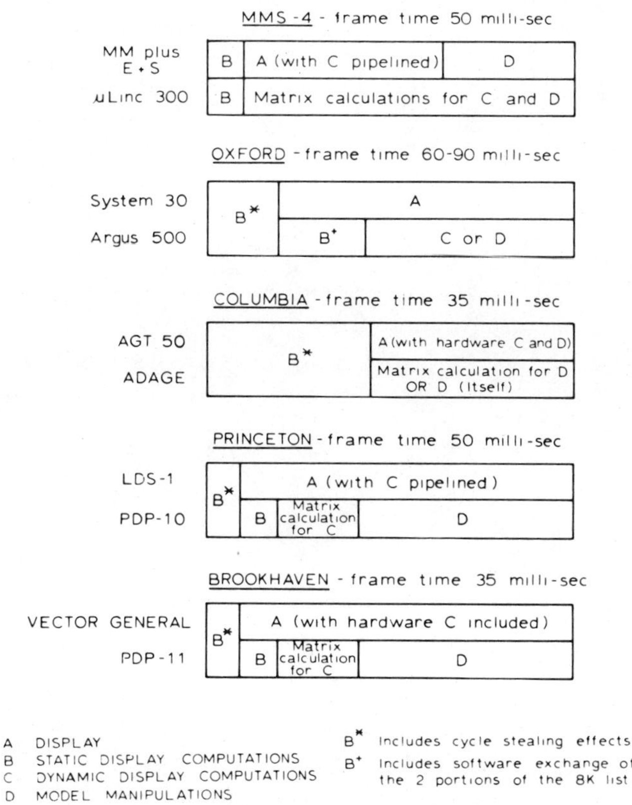

Figure 6. Comparative use of hardware as a fraction of the respective frame times.

TABLE 1. Summary of various molecular modeling systems[a]

Systems	MMS-4 (1971) (7, 9, 10)	Oxford (1970) (2)	Columbia[b] (1969) (13)	Princeton[c] (1970) (5, 14)	Brookhaven[d] (1973) (4, 19)
Main computer	Spear μ Linc-300	Ferranti Argus 500	Adage AGT 50	PDP-10	PDP 11/40
Core size	8K	15K	15K	64K	28K
Bits per word	12	24	30	36	16
Cycle time	1 μsec	2 μsec	1.5 μsec	1 μsec	450 nsec (effective)
Software matrix × vector[e]	400 μsec	320 μsec	250 μsec	100 μsec (approx)	< 100 μsec
Cost in $	≃ 300 K	≃ 150 K	≃ 250 K	≃ 500 K	< 100 K
Display					
Processor	See text	Ferranti System 30	AGT 50	E and S LDS-1	Vector General
Display List	No[f]	Yes	Yes	Yes	Yes (with 3-D hardware)
Subpicture	No	No	Yes	Potential not used	Potential; not yet integrated into system
Hardward matrix multiplier	Digital	No	Analog	Digital (18 bit)	Hybrid
Multiplier time	8 μsec	–	0	8 μsec	0 (effectively)
Frame time (typical)	≃ 50 msec	60–100 msec	≃ 35 msec	50 msec	30 msec
Words per vector (average)	10 for model, 4 for contours	2	3	3	3
Core avail to display (typical)	40K[g]	4K[h]	<15K	15K	8K
Static display					
Maximum number of vectors per frame[i]	4–6 K	4K[h]	5K	3K	3–5K
Limitations	Display	Core[h]	Core and display	Display	Core and display
Dynamic display					
Rotation type	Digital hardware	Software	Analog hardware	Digital hardware	Hardware
Effective time per vector	0 as pipelined with display	500 μsec	0 (analog)	0 as pipelined with display	0 (analog)

Maximum number of vectors rotated and displayed per frame[i]	4–6K	150	5K	3K	3–5K
Limitation	Display	Computation	Core and display	Display	Core and display
Model manipulations					
Maximum number[i] of vectors (atoms)	Hardware	Software		Software[j, k]	Software[j]
1. just for display	4K	200	4K-hardware	200	350
2. with transformed coordinates available	2K	200	150-software[l]	200	350
Summary					
Display of 3,500 vectors[i] with manipulation of 200 atom model					
1. just for display	< 1 frame	≃ 2 frames	< 1 frame	> 2 frames[j, k]	≃ 3/2 frames[j]
2. with transformed coordinates available for model	< 1 frame	≃ 2 frames	≃ 2 frames	> 2 frames	≃ 3/2 frames
Style	Stand alone	Stand alone; time shared with X-ray diffractometer	Used with high speed line to IBM 360/91 which creates the initial display files	Basically stand alone. PDP-10 is operated in a time-share mode	Basically stand alone though may be linked via telephone lines to data bases

[a] Reference 17 gives a broad analysis of other features of these and other systems. [b] Additional data supplied by Reidar Bornholdt. [c] Additional data supplied by Peter Bond. [d] Additional data supplied by Tom Koetzle and H. J. Bernstein. Part of the Crysnet System (4). [e] With translation. [f] Model display generated from transformed coordinates of molecule and connectivity list. Electron density map contours from a form similar to a display list (see text). [g] Macromodule processor includes 72K of 2 μsec 12-bit memories. [h] The system 30 display operates out of a 4K block of main memory. By software exchange of data in two 4K blocks an effective 8K list was obtained. [i] Considering typical vectors to be short, 1 to 2 inches. [j] Hardware option not yet implemented. [k] Coordinates generated from bond length and angle data rather than transformation of a previous conformation. [l] It is feasible to recover transformed coordinates after the analog rotation.

with macromodular registers to hold the external memory address and to buffer single data words. A similar facility is provided allowing the Linc access to the E and S equipment via a 72-bit register which holds commands and data, although general access to macromodular registers was considered inappropriate.

A separate routine, "display contours," has also been incorporated, providing for the display of electron density data as a three-dimensional network of lines (chicken wire representation) which is scaled, translated and oriented with the molecule (Fig. 1). Unlike the molecule display which uses coordinates and atomic connectivity, the contour data resemble a display list consisting of entries giving coordinates and a 2-bit code indicating whether the data should be interpreted as a "set point, draw to" or terminator command for the line generator.

Table 1 presents a summary of the characteristics of MMS-4 for comparison with four other systems. Although the figures can only be estimates for typical cases and the systems have appeared in response to diverse needs MMS-4 is seen to represent a viable approach. Its strong point lies in its performance when transformed coordinates of a model are required for calculation as well as display. This is in keeping with the potential to add further specialized function hardware to perform these calculations within one

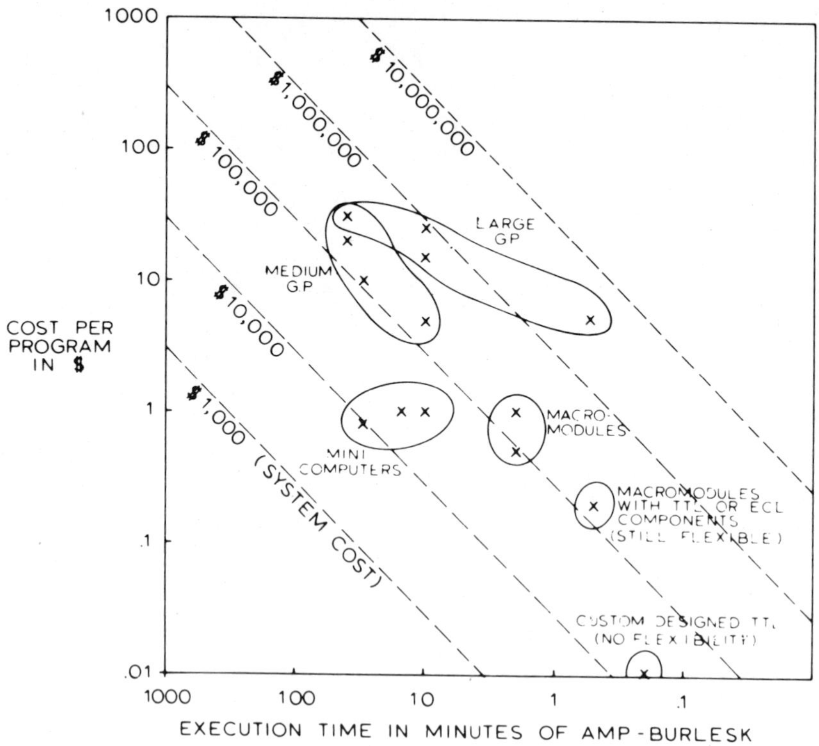

Figure 7. Cost versus performance plot for Burlesk operating on a typical molecule. The data plotted relate to the processing of N.M.R. data for Adenosine Monophosphate (1, 3).

or two frame times. To date this feature has only been used in monitoring interatomic distances and this can, because of the hardware coordinate generation, still be performed with the Linc. However, we anticipate other demands for this feature as molecular modeling becomes more sophisticated and we provide the human operator with more "senses" (real time calculations) to help evaluate conformations.

The trade-off associated with making the transformed coordinates available in MMS-4 involve extra memory, to hold two 24-bit copies of the coordinates and explicit molecule data lists, and display time because of the additional use of the hardware matrix multiplier. Figure 6 shows a breakdown of the use of the computer and display processor in terms of frame times for the systems included in Table 1 and indicates the trade-offs in terms of the major tasks (*A* to *D*) set out earlier.

While the preceding description of MMS-4 is less than complete it should indicate the "hand-crafted" approach that macromodules allow in system design, especially when performance is a concern.

Turning briefly to the other aspect of modeling—automatic refinement—automated electron density map fitting seems limited to investigation on large general purpose computers (8, 11). We have, however, been applying enumeration procedures to the conformation analysis of various forms of experimental data on small molecules. The procedure, which has become known as BURLESK (6), has been successfully applied to the structural determination of a series of mono- and dinucleotide phosphates from observation of lanthanide-induced shifts in the NMR spectra (1, 3). Identification of possible conformations of the peptide hormone angiotension II is also being actively pursued (18). To date this work has concentrated on the use of small laboratory computers such as the LINC, PDP-12 and Argus 500 which match the problem characteristics of small program size, modest input/output requirements but large demand for computation. However interest in larger molecules is starting to place these calculations outside reasonable limits and our attention has again turned to macromodules.

Coordinate generation is now even more a major concern as performance is no longer screened by frame times. It also differs from that used in the interactive modeling in that the number of atoms transformed by a single rotation matrix is always small. The pipeline form of the E and S matrix multiplier is now not as attractive (the 4 × 4 formulation is also wasteful) as when long streams of vectors could be processed. Control is more fragmented, operations are applied only to short data lists. While the MMS-4 system organization does not preclude such finer grain control it is not optimized for it. Alternative designs include providing in the macromodular machine some of the more straightforward program control functions, effectively increasing the grain size of the externally called routines, providing closer association between the macromodular routines and the programmable controller, perhaps more convenient memory access, or possibly both. As there is no reason for this to adversely affect the system performance for problems involving operations on long data lists, these alternatives also allow us to consider a unified system. Details of such an implementation of BURLESK are under active consideration.

Figure 7 presents data showing the cost versus performance characteristics of various implementations of BURLESK applied to the analysis of a typical molecule and again indicates the advantages offered by incorporation of specialized hardware.

In conclusion, our experience with MMS-4 and our analysis of other molecular modeling problems such as BUR-

LESK have confirmed that there are definite advantages to be gained in system performance by using the macromodular approach. The ability to provide specialized hardware easily gives the scientist new cost-performance options without requiring him to become a hardware expert.

In terms of an evolving system, MMS-4 has provided us with a flexible system with which to develop and evaluate new ideas. The high performance of the system allowed it to be used in a variety of projects involving both local and visiting scientists. The ensuing interactions lead us to believe that we have defined a set of primitive operations common to all computerized molecular modeling. These can be implemented as a set of easily callable, highly efficient routines using assembly language and special purpose hardware resulting in a molecule processor that can be used for modeling in much the same way as a floating point processor is used for arithmetic operations. By including display as one of the primitive operations and by taking advantages of the efficiencies of specialization this unit could act as the kernel of a modeling system whose cost to replicate is expected to be as low as $35,000. Development of this system, MMS-X, is currently being undertaken and should help make our style of molecular modeling available to a much larger community.

The development of MMS-4 has obviously been a team effort involving everyone associated with the Computer Systems Laboratory of Washington University. Our thanks are due especially to Charlie Molnar and Wes Clark who have, as directors of CSL, created macromodules for us to use and guided the development of MMS-4. Much of the software was written by John Fritsch and Tom Jacobi; without their help and ideas MMS-4 could not now exist. We also are grateful to a variety of people from near and far who have used the system and provided us with the criticism we need in the development of a more useful system. These include Scott Mathews, Len Banaszak, Larry Webb and Gil Brodsky from St. Louis and Jens Birktoft, Anders Liljas, Bob Kretsinger, Iain Paul and John and Sue Cutfield from elsewhere. The authors also thank David Shupe, Phil Lewis and Gayle Jerman for their help in preparing the manuscript.

REFERENCES

1. Barry, C. D., J. A. Glasel, R. J. P. Williams and A. V. Xavier. *J. Mol. Biol.* 84: 471, 1974.
2. Barry, C. D., and A. C. T. North. *Cold Spring Harbor Quant. Biol.* 36: 577, 1971.
3. Barry, C. D., A. C. T. North, J. A. Glasel, R. J. P. Williams and A. V. Xavier. *Nature* 232: 236, 1971.
4. Bernstein, H. J., et al. In: *Technical Proceedings of the 2nd Annual AEC Scientific Computing and Information Exchange Meeting.* Upton: Brookhaven National Laboratories, 1974, p. 148.
5. Bond, P. J. *Siggraph-ACM* 6: 13, 1972.
6. Bosshard, H. E., C. D. Barry, J. M. Fritsch, R. A. Ellis and G. R. Marshall. In: *Proceedings of 1972 Summer Simulation Conference.* LaJolla: Simulations Councils, 1972, p. 581.
7. Clark, W. A., and C. E. Molnar. In: *Computers in Biomedical Research,* edited by R. W. Stacy and B. R. Waxman. New York: Academic, 1974, vol. 4, p. 45.
8. Diamond, R. *J. Mol. Biol.* 82: 371, 1974.
9. Ellis, R. A. *Computer* 6: 28, 1973.
10. Fritsch, J. M., R. A. Ellis, T. H. Jacobi and G. R. Marshall. In: *Computers and Graphics.* New York: Pergamon. In press.
11. Greer, J. *J. Mol. Biol.* 82: 279, 1974.
12. Hill, E. D., D. Tsernoglou, L. Webb and L. J. Banaszak. *J. Mol. Biol.* 72: 577, 1972.
13. Katz, L., and C. Levinthal. *Ann. Rev. Biophys. Bioeng.* 1: 465, 1972.
14. Langridge, R. Abstract No. XXV-8 *Ninth International Congress of Crystallography,* Kyoto, Japan, 1972.
15. Levitt, M. *J. Mol. Biol.* 82: 279, 1974.
16. Marshall, G. R., J. Beitch, R. A. Ellis and J. M. Fritsch. *Diabetes* 21 (Suppl. 2): 506, 1972.
17. Marshall, G. R., H. E. Bosshard and R. A. Ellis. In: *Computer Representation and Manipulation of Chemical Information,* edited by W. T. Wipke, S. Heiler, R. Feldmann and E. Hyde. New York: Wiley, 1974, p. 203.
18. Marshall, G. R., H. E. Bosshard, J. D. Glickson, W. H. Vine and P. Needle-

MAN. In: *Recent Advances in Renal Physiology and Pharmacology,* edited by L. G. Wesson and G. M. Fanelli, Jr. Baltimore: University Park. 1974, p. 215.
19. MEYERS, E. F. *Nature* 232: 255, 1971.
20. ROBERTS, L. G. In: *The Computer Display Review.* Lexington, Mass: GMC Corp., 1972.
21. SCHERAGA, H. A. *Advan. Phys. Org. Chem.* 6: 106, 1968.
22. WARSHAL, A., AND S. LIFSON. *J. Chem. Phys.* 53: 582, 1970.

Toward a computerless computer facility[1]

THEODORE H. KEHL
University of Washington
Seattle, Washington 98195

Over the past 3 years we have begun to redirect a portion of our biomedical research facility's efforts toward the eventual goal of eliminating facility ownership of any computers. If completely successful our computer facility will consist of hardware personnel who will design and construct new pieces of equipment and software types who will design and construct new systems. None of our personnel will be devoted to maintenance of existing hardware nor to the task of running "user" programs for them. Rather our goal in this project is to build complete systems (both hardware and software) for our colleagues, install these systems in their laboratories, and assist in the further development of these systems. This paper will describe the rationale and procedure for this experiment. However, before going any further it should be emphasized that only a portion of our efforts are devoted to this experiment and we retain the option of continuing our existing centralized service. Hopefully, we will be able to make an honest, convincing comparison of the values in each approach.

PRIVATE, PERSONAL COMPUTERS

Probably the first private, personal computer was developed by Wes Clark and Charles Molnar. Because it was to be used in the laboratory it was given the name LINC—Laboratory Instrument Computer.[2] During the summer of 1963 some 80 scientists journeyed to Cambridge where they assembled some 40 LINC's. (These original 40 have become known as "Classic LINC's" to differentiate them from commercial versions that appeared later.) Using circuit boards and some prewired chassis these scientists, in two 1-month sessions, assembled the LINC's and learned all there was to know about "the care and feeding" of a LINC. Most did not know how to program—they were taught programming. Most knew little about electronics—they were taught electronics. At the end of this intensive training session each group packed their LINC into a moving van for shipment home.

Subsequently, each group did all of the programming and maintenance for their machine. (Professors Walter Woodbury and Albert Gordon of this Department were recipients of a Classic LINC and the author was able to observe their efforts firsthand.) Most groups expanded the memory and added teletypes. A few added other, more powerful, peripheral devices.

Meanwhile, Digital Equipment Corporation incorporated several LINC in-

[1] Supported by National Institutes of Health Grant #RR00374.

[2] Much of the early work was not formally published probably because its importance was not realized at the time. My knowledge comes from observing the LINC in J. W. Woodbury's laboratory and reviewing numerous grants involving LINC's.

novations (data break, LINC tapes, 12-bit words) into a computer—the PDP-5—which still later became the PDP-8. Thus the minicomputer revolution was begun.

Today, some 11 years later, more than 30 of the original 40 Classic LINC's are still operational under the original concept—private, personal computers with each group doing programming and maintenance. Thus the concept that biomedical scientists can and will maintain their own computer systems has been proven.

The point is that even though biomedical funding started the minicomputer revolution, it is relatively rare to find biomedical computer facilities that build computer hardware. In this paper we will briefly describe our design and construction techniques which allow us to build computer equipment. Perhaps more importantly, we will describe the administrative framework within which a group of computer scientists can develop these systems in close collaboration with basic biomedical scientists.

THE LOGIC MACHINE SYSTEM

Remarkable progress has been made in the past 10 years in the development of integrated circuits. Most of this progress has been accomplished, however, in the "data part" (or data path) of computer systems and less has been accomplished in the "control part." Conventional computer systems have "hardwired" control parts, but in an effort to have greater flexibility, microprogrammed computers have been built. In a microprogrammed computer the data part is controlled primarily by a microprogram, executing from a microprogram "engine," which replaces the hardwire control. It is usually thought that a microprogram is much easier to change than rewiring the hardwire part—in fact, depending on the architecture of the microprogram engine, rewiring may be much easier than remicroprogramming. Vertical microprogramming (for a more detailed description of vertical microprogramming see Husson's book (3)), however, provides the flexibility that allows major hardware changes simply by remicroprogramming. Then, if one groups MSI-LSI integrated circuits into "modules" (or functional units as we prefer to call them) one can construct digital devices such that the data part is accomplished by the functional units under a microprogrammed control part—in other words a machine in which the operations are specified by the microprogram—or, in still other words, a machine that does logic (not necessarily computation) and is therefore a logic machine.

Others, notably Clark and Molnar (2) and Bell et al. (1) have also developed modular computer systems. The Logic Machine differs from these earlier devices in using microprogrammed control part whereas the tendency in these earlier efforts was toward hardwired control.

To demonstrate the power of the Logic Machine, we have been able to construct a rather wide variety of computer equipment: graphics display terminals, floating point array processors, information retrieval systems, auxiliary processors for minicomputers and, indeed, minicomputers. Our minicomputer (4) is about as fast as conventionally designed systems but costs less to build and is far easier to maintain.

The point of this description of the Logic Machine is this: there has been a long history, going back to the LINC and continuing through macromodules, of computer hardware development within the biomedical framework. The reasons are many but some are concerned with the inability of computer vendors to ascertain correctly the needs of the biomedical community. Indeed, whenever biomedical scientists point out that current computer systems do not satisfy current needs, minicomputer manufacturers usually suggest that these same scientists do not understand the situation. If, for no other reason, we can develop a counterexample of hardware/software capabilities within the biomedical research community which demonstrates that computer science competence exists within the community and can be used to control some of the abuses of the vendors, our time is well spent. If, in addition and more importantly, it is possible to foster a "community of users" in which a group of scientists, working on similar research problems and using identical equipment, can mutually share hardware/software experiences, then we feel much of the barrier of communicating instrumentation methodology will be eliminated. This barrier exists in the current vendor/purchaser situation.

FACILITY DEVELOPMENT

Central to any successful computer facility is the staff. Without the staff the computer hardware is just boxes of junk. However, with a competent staff even rather marginal equipment can be made to perform useful computation. It is not possible to put enough stress on the importance of quality in a computer facility's staff. *People are everything: the hardware is just a tool.*

It follows, then, that the development of a computer facility is absolutely based on its staff. To develop a facility that has as a major goal the "total services" approach as described here requires specialists with a wide variety of skills. It has been our policy to allow all of our staff members to become as broadly experienced in as many different areas as possible. However, it is also expected that each will select an area of particular interest to him and in which he will develop a high degree of expertise.

To encourage this search for personnel concentration our policy is to assign a problem to a specific person (or two people at the most) and then allow that person to develop a solution to the problem. The important point is to allow the person to see the project all the way through. Usually the most often stated reason for dissatisfaction in employment comes from individuals feeling they are just "cogs" in machines—a particular problem for people working around computing machines. Allowing people to find out what they like to do and then permitting them to do complete tasks "makes for" happy employees. Happy employees, in turn, tend not to take other, better jobs.

The kind of employees one would seek to develop (and develop them you must) do not fall conveniently into the usual hardware or software category. That is to say they are not either programmers or engineers but, in fact, a little of both. Otherwise, in our microprogramming approach, the engineer would tend to blame all problems on the programmer and vice versa. These two individuals must work together—our policy is to have both individuals be the same person and thereby eliminate the arguments. Such individuals are difficult to find but, starting with someone with extensive systems programming experience, can be trained. A systems programming background is important because it emphasizes what computers do probably better than any other discipline. Electronics is not a very good background because the emphasis is too "atomic," i.e., an overview of how tasks are performed in computers is lacking when one studies just the circuits.

Given a firm grasp of "what computers do" from a systems programming viewpoint, it is relatively easy to move to microprogramming methodology and thence to functional unit design. Debugging designs is very important and, to support this hardware/software debugging, an automated system of computer device construction is highly desirable—probably mandatory. In our facility, we have developed both CAD (computer-aided design) software and semiautomatic wire-wrap equipment. This is described in detail in (5). Input to the software consists of quite abbreviated design specifications. A paper tape containing directives (commands) to the wire-wrap machine is the final output but, along the way, several kinds of listings are printed. With this system nearly all of the "scut" work of functional unit design is eliminated and, consequently, the designer can concentrate on the important problems.

Of course, without software the hardware construction capability is pointless. Our greatest asset in software development has been our good fortune in acquiring a microprogramm*able* (as opposed to microprogramm*ed*) computer—the Raytheon PB440. As a result, we have been able to construct special macroinstructions that, in turn, hastened the development of virtual memory operating systems. With these advanced systems we have been able to provide the hardware development support described above and, also, continue to provide the scientists who use our facility with quite good service. (It should be pointed out that our raison d'être is to provide service to biomedical scientists and what is described in this paper is only a small part of our total responsibility.)

As is the case with hardware development, our policy is to make one (or at most two) computer scientists responsible for any project. For example, two individuals completely developed our SUPER-BASIC—a compiler (rather than interpreter as are most BASIC systems) which runs in virtual memory. Currently commands are being added to SUPER-BASIC to allow us to use it as an SIL (system implementation language). Our overall goal is to do all of our system programming in SIL so that, given a special application, a user can modify his system, if necessary, to accommodate his needs without resorting to machine language programming.

CONCLUSIONS

It is now evident to many (if not most) that the large centralized institutional computer is not able to cope cost-effectively (and in some cases not at all) with many university computing problems, especially those involving physiological laboratory work. Networking may be an answer to some of these problems, but with the increasing sophistication of most users it may now be possible for computer facilities to provide more comprehensive but indirect service. An experiment in which a portion of our efforts are directed at building minicomputers and developing all the software for these systems is under way in our facility.

A major advantage of this approach is that when the inevitable error occurs, whether in hardware or software, a user of these services will be able to contact the computer scientist who made the original error in a matter of minutes.

REFERENCES

1. Bell, C. G., J. Grason and A. Newell. *Designing Computers and Digital Systems.* Maynard, Mass.: Digital Press, 1972.
2. Clark, W. H. Macromodular Computer Systems. *1967 Proc. Spring Joint Computer Conf.* 337–401, 1967.
3. Husson, S. S. *Microprogramming Principles and Practices.* Englewood Cliffs, N. J.: Prentice-Hall, 1970.
4. Kehl, T. H., L. Dunkel and C. Moss. LM²—A logic machine mini-computer. *IEEE Trans. on Computers.* In press.
5. Kehl, T. H., C. Moss and L. Dunkel. Design automation for custom hardware in computer research. *IEEE Trans. Education.* 168–170, Aug. 1974.

Session III

Computer equipment configurations

Rapporteur's summary

HAROLD SHIPTON
University of Iowa
Iowa City, Iowa 52242

I will speak from an engineering point of view, and I shall try to address my remarks to the users to some extent rather than to the people who are handling the problems on a day-to-day basis.

One of the things that came through each of the five papers was an indication that special-purpose systems are coming back into vogue, even if not all of us are going to follow Ted Kehl's habit of rolling his own, and for those of you who may wish to roll your own, I should tell you that you can get hooked very easily on habits of that kind so do be careful.

Engineers are, hopefully, visionary people and therefore quite a lot of what was said was in the nature of prophecies. And not all prophecies come true. So I think you should recognize that it will be some time before macromodules replace many machines, although I believe that the trend is there and I think it will indeed happen. But just to spend a moment on the question of prophecies. Remember that some of those that did not come true seemed, at the time, to be very exciting, and I put this in as a caution to the user rather than the doer group. A few years ago when we went to a computer science meeting you heard all about cryogenics. We were going to cool our computers and move towards an energyless computer system. It didn't happen.

I was surprised to find that not a word was said in this session about memories. Memory technology is changing very rapidly. Dr. Prager saved the situation by one of her explanatory slides and she did say, "the laser memory is progressing well and the electronic components are all at hand." I was glad to hear it because I think this is an important technology. It is worth our while thinking about what the impact will be of a compact, inexpensive, small sized and rapid-access memory. The technologies are here, or just about here, and I think we shall be under the impact of new memory technology before the next of these Conferences. It will be very important to get rid of a lot of mechanical devices for they have been the stumbling block for many systems. I think such devices will be compact, easily used and just within our means, with the help of the NIH.

There is one impressive aspect in each of the presentations. Computers are beginning to learn from biology. Twenty-five years ago it was fashionable to make analogies between the digital computer and the brain and I suspect the naiveté of some of those analogies turned off many good promising computerniks. Now we are doing the sensible thing. We are making the computer more like a living system. We coupled it directly to the external world. We used preprocessing as does our nervous system to simplify and facilitate transportation to the central station. We are beginning to put our redundancy systems out where they can do good. Then we are moving towards a synchronous independent parallel processor. The serial, one thing at a time, Von Neumann machine was abandoned by God a long time ago. Lastly, we are beginning to put our output in a natural pictorial from to which we are well geared as biological entities. I suggest that the message that we should take from the five papers in the Engineering Session is that in serving the life sciences the engineer is in a position to learn from them as well.

Session IV

Data base systems

Introductory remarks

JOHN A. STARKWEATHER, *Organizer and Chairperson*
University of California
San Francisco, California 94143

Although many parts of a computer system can vary in size, it is the amount of data which can be stored that more often than not determines whether we call it a large or a small system. Consider the smallest systems, such as pocket calculators, which have only a few registers of data memory. It is predictable that their capability will grow in the direction of increased data storage.

The typewriter-sized Datapoint Terminal in the lobby is a complete computer which can access data stored on small Philips cassettes just like the ones used for audio recording of voice. I have been developing an instructional use of this equipment which is likely to grow in the direction of improved access to data. When I connect this machine to larger equipment, it is more often in order to access more data than to gain more computing power. That is, of course, related to the nature of the problem.

We are all aware that a logical separation can be made between the data that we wish to manage or manipulate and the programs that do the manipulation.

Once upon a time, a few computer generations ago, it used to be considered clever to blur this distinction and to manipulate *program* elements in the course of program execution.

The fact is that the computer itself makes no distinction between data and programs in its memory and will not balk at manipulating program elements or indeed trying to execute data as instructions. But these things cause trouble among humans, and modern programmers will not usually admit to such evil practices. Their data are separate, pure, and structured.

This morning we will hear about systems that provide access to large amounts of data. These are necessarily large systems and team efforts lie behind their development. Our speakers are concerned about overall function and how a system serves its users rather than internal technical details. We will hear about examples from industrial and medical research, from two different settings of clinical patient record handling, and from a production center for bibliographic information.

Functional goals and problems in large-scale patient record management and automated screening

MORRIS F. COLLEN, LOU S. DAVIS,
EDMUND E. VAN BRUNT AND JOSEPH F. TERDIMAN
Permanente Medical Group and
Kaiser Foundation Research Institute
Oakland, California 94611

Kaiser-Permanente Medical Care Program, the largest group practice, prepaid health plan in the United States, delivers comprehensive hospital and medical services to more than one million persons in the San Francisco-Bay area. In this area, it currently owns and operates 11 hospitals and 15 outpatient medical office facilities for more than 1,000 physicians.

In studying the costs of our group medical practice, it was determined that 80% of physician services are for outpatient care and only 20% for hospital care. Approximately one-fourth of all visits to the clinics are to the department of medicine; and more than one-third of all new patients requesting appointments to internists desire a health checkup. In the hospital services, it was determined that about one-third of the daily hospital costs were for information processing (reporting tests, communicating orders, and the like). Accordingly, we directed computer applications to these areas, with first priority in the outpatient department to the health checkup area, and in the hospital, to the information-handling problem.

AUTOMATED SCREENING

The greatest impact of computers in our medical practice was by utilizing automated multiphasic health screening programs for periodic health examinations to ambulatory patients.

Automated multiphasic testing is a systems approach to periodic health examinations which conserves the physician's time by relieving him of most of the routine repetitive data acquisition aspects of the health checkup (1, 2, 7–9). Automated multitest laboratories with automated, electronic and computer equipment are being used increasingly as an integral part of a routine periodic health examination. There are now 200 such programs registered in the USA, 20 in Europe, 5 in Japan, and 2 in Australia.

The automated multiphasic program generally provides electrocardiography, anthropometry, chest and breast X-rays, visual acuity tests, tonometry, audiometry, respirometry, an automated health questionnaire, and clinical laboratory tests including a panel of blood chemistries. Before the patient leaves the multitest laboratory, additional indicated procedures can be arranged in accordance with computer "advice" rules programmed by criteria which have been predetermined by the medical staff. When all test reports are received, the computer prints out a summary report.

When the physician sees the patient, he reviews the summary report; and then needs only to direct his attention toward elaborating on those questions to which the patient has answered "yes," and to the test abnormalities reported from the automated multitest laboratories. He then completes the physical examination, makes his diagnosis, and then can arrange any necessary follow-up care in the usual manner.

In our program, if the patient is agreeable, he may have an immediate physical examination by a trained nurse working under the supervision of a physician. Referrals to health education and preventive maintenance clinics are then arranged as indicated.

Our 10 years of experience with the computerized multiphasic health check-up have shown that:

1) Automated equipment improves efficiency, since the speed, accuracy and economy of testing are greatly improved. Four to five times as many tests can be provided to the patient for the same cost. Our direct costs per patient tested in our Oakland multiphasic laboratory are $14.50 ($12.00 for direct operational costs and $2.50 for data processing).

2) Patient's time is saved, since several times as many tests can be provided per hour.

3) The productivity of physicians is greatly improved when they provide periodic health examinations to patients who have received prior testing by automated multiphasic laboratories. Where automated multiphasic testing has been completed in advance of the physician's examination, it has been our experience that 50 to 70% of patients can complete their "checkup" with one visit to the physician. With the addition of computer "advice rules" which automatically propose necessary retesting and additional procedures, our studies show that 80 to 85% of patients can be taken care of by one physician visit. Privately operated automated multiphasic health testing laboratories in this country have generally reported an increase in physician productivity of at least 50%.

4) Comparative cost analyses of ambulatory care services received over a 12-month period following health checkups by multiphasic testing as compared to the traditional mode showed a decrease in costs of approximately 20%.

5) A controlled long-term study has revealed that middle-aged men, after 7 years of having regular health checkups, show a significant reduction in self-rated disability, a greater proportion are working, and they have greater earnings over the 7-year period (5).

6) Multiphasic testing is an effective method of providing medical checkups. Based on our experience with a representative middle-class population, it can be expected that one-third of all adults (and two-thirds of those over age 45) will have some clinically important abnormality reported by their physicians. One-fourth will have a significant abnormality reported on chest X-ray and one-fifth on electrocardiogram. Eight percent of all adults will have hypertension. Five percent of women will have anemia. Two and one-half percent of all adults will have diabetes mellitus; 2% will have deafness and 1% glaucoma. One in every 500 women over age 50 will have breast cancer, and one in every 5,000 adults will have a parathyroid tumor.

7) Multiphasic health testing is very acceptable to patients (14). A great majority of persons surveyed have expressed a high degree of satisfaction.

8) Computer data processing with a continuing patient computer file provides to the physician in a timely manner, a comprehensive profile of the patient's current and prior status. It allows efficient handling of large volumes of medical information on great numbers of patients over long periods of time. It permits on-line quality control, thereby decreasing rates of errors due to instrumental variations, personnel errors and insertion of data into wrong records.

The rapid and widespread acceptance of automated multiphasic health checkups in the United States indicates that in the near future it is likely that every large health care program will have a multiphasic health testing laboratory that will be available to patients for periodic health examinations, for health evaluations for special purposes (such as employment and insurance examinations), and for early sickness consultations, and diagnostic surveys. The multiphasic laboratories also will be used for admission and preoperative examinations for ambulatory patients going into the hospital.

New entry systems to medical care are now being tested and evaluated that use computerized multiphasic health testing services to provide a new entry mode to direct and refer the sick and the well for appropriate care, including health care for the well, and preventive maintenance services for the treatment of early and chronic disease (13). Computer-directed protocols are evolving for guidance of nurse practitioners who, under physician supervision, provide examinations and monitor the management of patients with common chronic conditions (hypertension, diabetes, obesity, and so on).

PATIENT RECORD MANAGEMENT

Outpatient clinic data systems

Several large medical care programs in the USA are developing computer data collection systems for their office patients. Simple structured "encounter" forms are generally being used for the recording of essential medical information by physicians since direct interactive computer terminals are still too expensive, and the computer processing of natural language prose reports is very difficult.

In our San Francisco Permanente Medical Group offices, patient registration data and physicians' diagnoses for 2,000 outpatient visits daily were recorded on specially designed standardized forms. Between 90 and 95% of the diagnoses were entered by the physicians from a structured format. The remaining unstructured data were manually inscribed on the form; thereby, less than 10% of this natural language data to be stored required keypunching for input. Using standardized forms wherever possible, with consistent terminology for data input to medical records, simplified the storage and retrieval of medical data. The data from these forms were entered by on-line electric typewriters directly into each patient's computer record (15).

In the outpatient pharmacy, the data from 1,200 prescriptions daily (including patient and physician identification, drug name and dose, prescription refill and drug usage dates) were entered by pharmacists using on-line electric typewriters, directly into the appropriate patients' computer medical records. These records resided in direct-access mode in the central facility. Container labels were produced under program control, to be dispensed with the drug.

A clinical laboratory computer system was being implemented which could handle data for 3,000 tests a day. An on-line electric typewriter in the emergency clinic could retrieve at any time the above outpatient clinic data from the patients' computer medical record.

Hospital care

Since about one-third of a hospital's per diem costs are for information handling, the greatest impact of computer technology in the hospital will be on its medical information system (4). Such a hospital information computer system will eventually provide the following functions:

1) Communicate individual patient data from the professionals providing medical care (doctors, nurses, technicians, etc.), into the patient's computer medical record; and then to other professionals (e.g., dietitian) and hospital services (e.g., radiology) (3).

2) Communicate patient data from subsystem components (e.g., automated multiphasic screening laboratory or intensive care unit) into the patient's computer medical record.

3) Communicate between clinical services (i.e., nursing stations) to ancillary services (e.g., ECG or radiology).

4) Establish scheduling files and communicate information for scheduling of patients, personnel and medical care services.

5) Establish a data base for administrative and business functions.

6) Establish a medical data base that can fulfill research and teaching objectives.

Such comprehensive computer hospital information systems are not yet operational anywhere in the world. However, in large medical centers some subsystem components, such as automated multiphasic health testing programs, patient monitoring systems, admission and bed census systems, outpatient scheduling systems, and pilot hospital information systems are in various stages of development.

Such a pilot medical data system was tested within our San Francisco Kaiser Foundation Hospital on the Pediatric Service. The hospital data subsystem consisted of a satellite computer linked by telephone line with the central computer facility. The satellite processor drove 24 visual display terminals, with associated "light-pen" sensor, and an electronic printer enclosed in a sound-proofing box.

The system of terminals was evaluated by pediatric doctors and nurses for data input and output of diagnoses, physicians' orders (including orders for drugs and general nursing orders), drug administration data, medication schedules, and clinical laboratory results.

The patient computer medical record

A computer-stored, integrated medical record was essential to the functioning of a computerized medical data system for a large medical program. Such a record, and the system that maintains it, must be so structured as to store all classes of patient-related data: identification, administrative, and medical, inpatient and outpatient. It must be capable of continual updating over the lifetime of the patient, accepting the variable format of data input, responding to the need for real-time inquiry to individual patient records, and it must be maintained by reliable error detection and recovery routines.

The file organization of our data files was based on an integrated, continuous, variable length, variable format record. Our data base contained more than one million patients' records. All records were individually retrievable by each patient's medical record number on a direct access basis (10, 11).

Over the past 5 years, the pilot medical data system described above was developed and tested and financially supported primarily from federal research funds.

TABLE 1. Regional medical services computer center approximate costs (dollars)

	Annual total	Per member per year[a]	Per doctor office visit	Per hospital day
Staff/services	500,000	0.50	0.06	0.50
Computing equipment	500,000	0.50	0.12	1.00
Data base storage	250,000	0.25	0.09	0.25
Total	1,250,000	1.25	0.27	1.75

[a] For 1.15 million members.

TABLE 2. Approximate unit medical EDP costs (dollars)

Application	Annual total	Unit cost
Multiphasic screening	50,000	2.50/patient
Clinical laboratory	50,000	.06/test
Pharmacy	15,000	.05/prescription
Hospital admissions	10,000	1.00/admission
Nursing station system	500,000	
Essential medical information	5,000/terminal	.03/retrieval[a]
Doctor office visits	90,000	.17/visit
Electrocardiography	40,000[b]	1.35/ECG

[a] Based on one series of inquiries per non-appointment visit (Emergency Room and "walk-in" ambulatory clinics). [b] In groups of four facilities.

These research support funds were discontinued the end of last year.

As a result of the increasing capital cost requirements associated with a large central computer system, an organizational decision was then made to embark on a program of using multiple modules employing lower cost minicomputers for medical applications. Accordingly, separate small processor systems are being developed and installed for an automated multiphasic health testing (AMHT) system, an ECG automated analysis system, and an outpatient pharmacy system. It is anticipated that if these three modular subsystems can be successfully integrated by a minicomputer data base management system, then other clinical application modules (outpatient visit-encounters, radiology, surgical pathology, and so on) will follow.

Costs

By the end of last year there had been sufficient experience with testing and operating the patient record management system that the costs were identifiable for the data base system and for the individual applications (16).

Table 1 lists the approximate costs for staff services, computing equipment, and data base storage for the regional medical services computer center. Annual total costs, and the fraction per member per year, are based on equipment configuration and storage capacity sufficient to meet the needs of four medical centers of the size of the pilot facility. Estimates of cost required for expansion to support the data processing needs of 10 medical centers providing services to one million patients are reflected in the incremental increases shown in the "per doctor office visit" and "per hospital day" columns.

Costs of "applications" in the pilot facility, listed as approximate unit medical electronic data processing costs, are shown in Table 2.

Galbraith (12) was one of the first to point out the problems and the consequences of such large complex technological systems, namely: *1*) long-term investment—the more complex the technology the greater the lead time from onset of planning to completion of the system; *2*) heavy investment of capital equipment and manpower; *3*) need for accurate planning to inflexibly commit people and capital for long lead times due to usually several years between planning and completion; *4*) specialized technology manpower, to design, develop and implement most effectively; and *5*) organization—especially competent to coordinate mixed medical, systems and engineering specialists.

For a medical data base management system, the situation is further aggravated by the lack of standardization of medical practice and the difficulties in demonstrating cost-effectiveness. In 2 to 3 years, we expect to have sufficient data for a meaningful comparison of the integrated modular approach to the large central computer approach, taking into account the many factors that must be considered when evaluating cost-effectiveness.

REFERENCES

1. Collen, M. F. The multitest laboratory in health care of the future. *Hospitals* 41: 119, 1967.
2. Collen, M. F. Automated multiphasic screening. In: *Presymptomatic Detection and Early Diagnosis*, edited by C. Sharp and H. Keen. London: Pitman, 1968, chapt. 2, p. 25–66.
3. Collen, M. F. General requirements for a medical information system. *Comput. Biomed. Res.* 3: 393, 1970.
4. Collen, M. F., editor. *Hospital Computer Systems*. New York: Wiley, 1974.
5. Collen, M. F., L. G. Dales, G. D. Friedman, C. D. Flagle, R. Feldman, and A. Siegelaub. Multiphasic checkup evaluation study. 4. Preliminary cost benefit analysis for middle aged males. *Prev. Med.* 2: 236, 1973.
6. Collen, M. F., and L. S. Davis. Computerized medical records in multiphasic testing. *Computer* 6: 23, 1973.
7. Collen, M. F., R. Feldman, J. Barbaccia, J. Dunn, R. Greenblatt and A. Mather. *Provisional Guidelines for Automated Multiphasic Testing and Services. Volume 2. Operational Manual.* Washington, D.C.: U. S. Govt. Printing Office, 1970.
8. Collen, M. F., R. Feldman, A. B. Siegelaub et al. Dollar cost per positive test for automated multiphasic screening. *New Engl. J. Med.* 283: 459, 1970.
9. Collen M. F., L. Rubin and L. S. Davis. Computers in multiphasic screening. In: *Computers in Biomedical Research*, edited by R. W. Stacy and B. D. Waxman. New York: Academic, 1965, vol. 1 chapt. 14.
10. Davis, L. S. Patient computer medical records. *Comput. Biomed. Res.* 3: 539, 1970.
11. Davis, L. S. A system approach to medical information. *Method. Inform. Med.* 12: 1, 1973.
12. Galbraith, J. K. The imperatives of technology. In: *The New Industrial State*. Boston: Houghton Mifflin, 1967.
13. Garfield, S. R. The delivery of medical care. *Sci. Am.* 222: 15, 1970.
14. Soghikian, K., and F. B. Collen. Acceptance of multiphasic screening examinations by patients. *Bull. New York Acad. Med. 2nd Ser.*, 45/12: 1366, 1969.
15. Van Brunt, E. E. The Kaiser-Permanente medical information system. *Comput. Biomed. Res.* 3: 477, 1970.
16. Van Brunt, E. E., L. S. Davis, J. F. Terdiman and M. F. Collen. The Kaiser-Permanente pilot medical data system. *Proc. Intern. Conf. Health Technol. Systems*. San Francisco, 1973.

Use of a psychiatric patient record system

BERNARD C. GLUECK, R. PETER ERICSON
AND CHARLES F. STROEBEL
Institute of Living
Hartford, Connecticut 06106

We began our efforts to computerize all of the functions of the Institute of Living, a 400 bed, private, nonprofit psychiatric hospital, including the full patient record, in 1962. In the intervening years we have gradually developed our concepts of appropriate computer and terminal hardware and the kind of programming capability and file structure necessary for optimum utilization of the system for day-to-day clinical care of our patients, as well as furnishing support for the administrative and business operations of the institution.

The research uses of the system also are a major consideration, since the system was developed by the Research Department, a somewhat unique point of approach as compared to most other hospital computer systems. While none of the presently available hardware and software is completely satisfactory, we feel we are currently able to do a very competent job with the equipment and programming capabilities available.

Figure 1 is a schematic representation of the computer system that is currently in full-time use at the Institute. This system has been operational since the middle of January of this year. It consists of two Digital Equipment Corporation PDP-15 computers with tape drives, disc drives, and so on. The two systems are basically identical, except for a card reader and a line printer on the secondary system. Both systems have 73,738 18-bit, 800-nsec core memory, with 10.24 million word storage on each of the disc packs. The primary system uses a multiplexor that will handle 40 lines, while the secondary system has 16 channels. This represents a rather sizable computer system for a relatively small hospital.

We were forced to make a decision to continue development of a system completely within the hospital, rather than using a larger, shared system with other medical facilities some 4 years ago, because of major difficulties in our attempts to develop a large-scale, shared hospital data processing facility. We felt that we could afford a duplexed system to give us the necessary 24-hour availability and reliability in part, because of the improvement in price/performance ratio provided by the third-generation computer systems, as represented in the PDP-15.

A second critical factor was the availability of the "MUMPS" programming language, which was specifically developed for the medical environment, and appeared to be working quite satisfactorily in a number of applications at Massachusetts General Hospital (1, 2).

We have had the usual problems in installing, and making operational, a new hardware and software system, compounded by the time and effort spent in

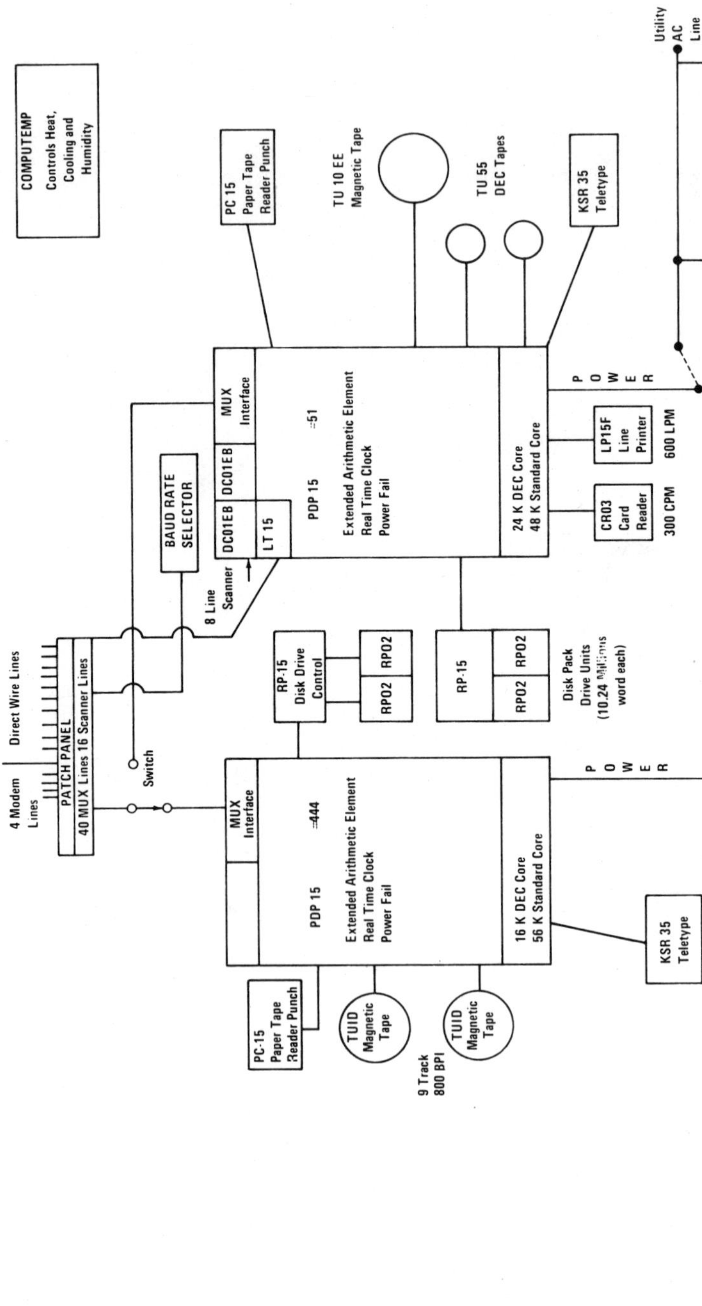

Figure 1. The Institute of Living's PDP-15 Time Sharing System.

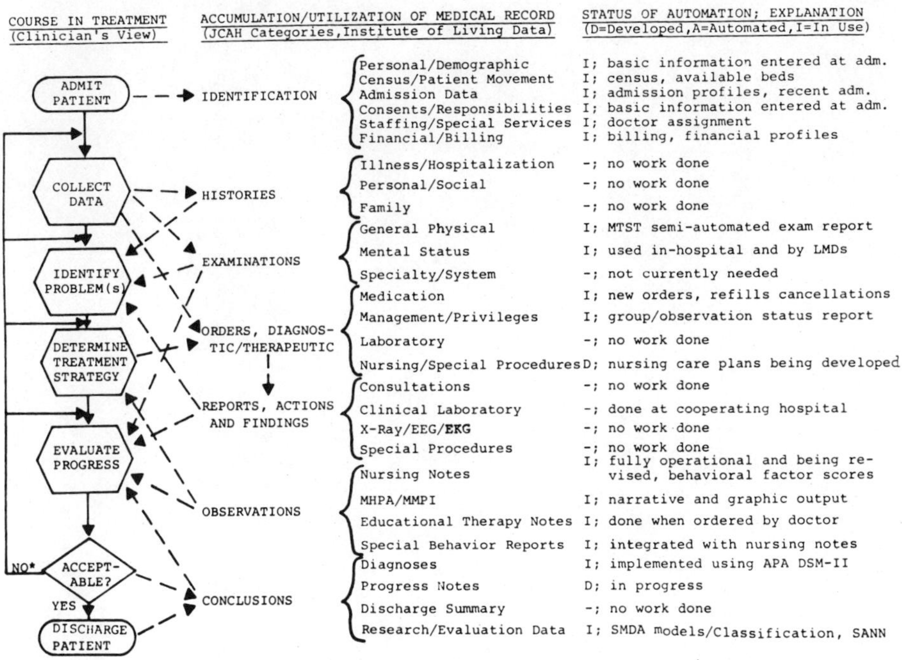

Figure 2. Management of the patient via management of his clinical data—a status report.

attempting the development of yet another programming system called "HEALTH." HEALTH was an attempt to develop a generalized data base management system that would eliminate the necessity for specific applications programs as such. I still am of the opinion that this is the proper direction in which we should be moving, but we suffered from limitations of funding and adequate manpower to complete the rather enormous task posed by the development of such a system. With curtailment on federal funding that has occurred in the past 2 years, we were finally forced to call a halt to work on the HEALTH system and return to the basic MUMPS data base structure.

Our overall plan for the development of the full computer system is shown in Fig. 2. Of particular importance is the middle column which indicates subsystems that are organized according to the Joint Commission on Accreditation of Hospitals standards. We have tried to develop the patient data base in as close conformity to these standards as possible, so that there will be relatively minor problems in getting Joint Commission acceptance of the system once it is fully operational. The right-hand column shows the status of the various applications, as well as the rather significant amount of work still to be done.

One practical note concerns system maintanance, both hardware and software. We have had more than the usual number of problems in maintaining the system's hardware, occasioned in part by the complexities of the system, in part by problems that all of the computer manufacturers seem to have these days in terms of adequate field service personnel, and adequate field inventories of replacement parts, and in part by using equipment from several vendors. Main-

TABLE 1. PLIS items for resident patient files

Item name	*Maintenance method (transaction)*
Administrative Items	
Case number	Admission
Year number	"
Admission date	"
Admission to IOL (first, readm., etc.)	"
Legal status	"
Permanent doctor	Admission and dr. assignment
Current doctor (= perm. dr. if perm. assignment)	" " "
Date of doctor assignment	" " "
Future doctor	Doctor assignment
Date perm. doctor due back	" "
Psychologist name (if any)	Assign psychologist
Consulting doctor and place	Consultation
Mail (restricted or not)	Mail restriction change
Date of latest change of mail	" " "
Signed voluntary slip (yes, no)	Front sheet input
Personal Items	
Title	Admission
Last name	"
First name and middle initial	"
Sex	"
Marital status	
If female and married	"
husband's name	"
maiden last name	"
maiden first name and middle initial	"
Birthdate	"
Religion	"
Name of responsible parties	Front sheet input
Relationship of responsible parties	" " "
Address of responsible parties	" " "
Phone number of responsible parties	" " "
State or country of patient	" " "
If from Connecticut, name of town	
Occupation	" " "
Race	" " "
Education (highest level)	" " "
Clinical Data	
Medication data	Pharmacy system
MMPI data	MMPI system
Patient Movement Items	
Permanent unit and room	Admission and transfer
Current unit and room	
(= perm. unit and room if perm. assgn.)	" "
Transfer date	" "
Unit admission date	" "
Group	" "
Group change date	" "
Observation type (close, const., etc.)	" "
Observation change date	" "
Current privilege cards	Privilege change
Dates cards were issued	" "
Residence status (resident, V-RR, etc.)	Admission, visit, return, discharge
If planned departure, on visit-RR, AWOL:	
departure type	Planned departure or departure
room reserved (yes, no)	" " "
departure date	" " "
return date	" " "
accompanied by	" " "

If planned discharge or discharge	
discharge type	Planned departure or discharge
discharge date	" " "
return date (if any)	" " "
Medical Items	
For admission diagnostic impression	
diagnosis and additional comments	Diagnosis input
Physical illness and add. comments	" "
Date of diagnosis	" "
Admitting doctor	" "
For staffing and review and discharge diagnoses:	
diagnosis	" "
Date of diagnosis	" "
Research Items	
Attributes:	
(SANN, TM, Alpha, Match, ADMSERV*)	Attributes update

tenance of the software has been complicated because the MUMPS language is still in the process of evolution, and there are several different levels of complexity available. One happy note here is the attempt, currently underway as a joint effort of the National Bureau of Standards and the National Institute of Mental Health, to try to standardize the various dialects of MUMPS, at least seven at last count, and come up with a set of standards that will apply to all MUMPS users and systems. This certainly would facilitate interchange of application programs and techniques among the various hospital and medical users of the MUMPS system.

The items of information that are currently available in the Patient Logistic Information System (PLIS) are shown in Table 1. The item starred, ADMSERV, was recently added to the current file. We

```
THIS PROGRAM WILL SEARCH FILES & PRINT ITEMS
DIAG ? N
MEDS ? -
PRINT FIELDS ?

TITLE = ADMISSION SERVICE

PROG RPC

SEARCH INPATIENTS ? Y
DAY PATIENTS ? Y
DISCHARGES ? Y

OUTPUT DEVICE (H,T,S,NO.): S
YOU HAVE SPOOLER NO. 1    YOUR JOB NUMBER IS 2

SEND TO AND DESCR.: P ERICSON

MAX PAGES IF  10:

UNLOCK THIS DEVICE? Y
JOB EXECUTION STARTED.
THANK YOU
```

Figure 3. Extractor program for ADMSERV.

would like to describe the reason why this item was added and demonstrate the ease with which the file can be expanded using this example.

One of our doctors requested a list of all patients who had gone through the Admission Service since its inception 18 months ago. He also requested specific data on each patient. Our PLIS system did not distinguish these patients from the rest of the hospital population. As a result, their records could not be found and searched for the information requested. In addition, no report existed that displayed the requested data.

A new attribute (ADMSERV) was added to the system by means of a terminal transaction requiring about 5 minutes of effort. All past patients (entering the hospital through the admission service) were coded by a clerk at a terminal (20 min). The extractor program was modified to produce the requested data (20 min) (Fig. 3). The requested report was produced (15 min) (Fig. 4). Elapsed time from request to report was one day.

A second example of the flexibility of the system for research purposes is illustrated by the following situation: In an effort to evaluate the treatment of alcoholism at the Institute of Living, two of the psychiatrists requested a list of all patients in the hospital who had a problem with misuse of alcohol. The general extractor program was used to search the files for specifically identified diagnoses, and print a list of matches (Fig. 5).

Since alcohol abuse may occur as a symptom in several psychiatric illnesses, multiple diagnoses were given for the

```
ADMISSION SERVICE   WEDNESDAY 5/8/74   3:27 PM

PATIENT # AND NAME      BRD1   CAMERON, I.          67       12/18/73
(1)731218*300.4*301.81   (2)740103*300.4*301.4*309.3    (3)   (4)

PATIENT # AND NAME      BR 2   IGER, H.             23       10/11/73
(1)731011*295.3     (2)731029*295.3*309.2      (3)     (4)

PATIENT # AND NAME      BU 2   ZELLER, W.           19      01/02/74
(1)740102*295.0*R295.3*R295.2*R295.7*R310.0  (2)740117*295.3   (3)  (4)
BRIGHAM 2

PATIENT # AND NAME      ESSX   DAVIS, C.            37      03/19/74
(1)740319*300.4*R296.1*R295.5*R298.0  (2)740404*300.4*301.81   (3)  (4)
BRIGHAM 1

PATIENT # AND NAME      FU 1   PENTZ, P.            18      01/16/74
(1)740116*300.4*304.7*304.2*304.5*R295.3*301.5  (2)740201*295.5  (3)  (4)
TODD 1

PATIENT # AND NAME      ST 1   IGER, H.             20       06/12/73
(1)730612*295.3*R295.74   (2)730626*300.4*301.2   (3)     (4)
BRIGHAM 2

PATIENT # AND NAME      TH 2   ROBBINS, J.          23       03/07/74
(1)740307*295.3     (2)740321*295.3       (3)        (4)
BRIGHAM 2

PATIENT # AND NAME      ST 1   MUCHA, T.            61       12/22/73
(1)731222*296.33*R295.7    (2)740107*291.5*R293.3*R294.1*R293.0*303.2  (3)(4)
BRIGHAM 2

PATIENT # AND NAME      TH 2   SILVERMAN, S.        20      07/19/73
(1)730719*304.5*304.4*301.77   (2)730803*301.77*304.4*304.5*304.6    (3)    (4)
BUTLER 2

PATIENT # AND NAME      TH 2   MILLER, L.           25      03/12/74
(1)740312*295.24*R295.3   (2)740329*295.2      (3)     (4)
```

Figure 4. Output from extractor, ADMSERV.

```
I.O.L. MUMPS-P
PW  CALL XMB:
ID :
_C RPZ

THIS PROGRAM WILL SEARCH FILES & PRINT ITEMS
DIAG ? Y

ENTER DIAG TO BE MATCHED
DIAG 1 = 291 (ALCOHOL PSYCHOSIS)
DIAG 2 = 309.13 (NON-PSYCHOTIC OBS, ALCOHOL)
DIAG 3 = 301.81 (PASSIVE-AGGRESSIVE PERSONALITY)
DIAG 4 = 303 (ALCOHOLISM)
DIAG 5 =
MEDS ?
PRINT FIELDS ?

TITLE = ALCOHOLICS

PROG ALREADY AVAILABLE

SEARCH INPATIENTS ? Y
DAY PATIENTS ? Y
DISCHARGES ? Y

OUTPUT DEVICE (H,T,S,NO.): S
YOU HAVE SPOOLER NO. 2   YOUR JOB NUMBER IS 3

SEND TO AND DESCR.: P ERICSON

MAX PAGES IF   10:

UNLOCK THIS DEVICE? Y
JOB EXECUTION STARTED.
THANK YOU
```

Figure 5. Extractor program, alcoholics.

search. The flexibility of the extractor is evident by its ability to search for a general class, first three digits, or for a sub-type, five digits, in the same operation. This flexibility is one of the real efficiencies available in MUMPS (Fig. 6).

We have included the two examples above to demonstrate the ease and practicality of use of our PDP-15 MUMPS installation. In the time remaining, we would like to describe some of the uses that have already been made of our patient data base. One of our most useful applications has been the conversion of the Nursing Notes to the automated system. This has been very successful, both from the standpoint of enthusiastic acceptance by the nursing personnel, who see themselves doing a far superior job of reporting in a much more efficient fashion, and also by the clinical staff, who have found the information generated by the Nursing Note to be of considerable importance in following the progress of their patients. We have tried to make this information readily available by providing graphic displays on the video terminals of all of the nursing factor scores, which are updated daily. In addition to this graphic display, we have been utilizing a sequential analysis program to determine when significant change in the various factor scores had occurred. The information obtained from this program has also been utilized to help in the prediction of disruptive behavior on the wards, potential for running away from the hospital, response to the various therapeutic interventions, and the like.

A second major area of current research is the development of statistical models for optimum classification of patients for various purposes, e.g., diagnostic groupings, optimum treatment response, picking matched cohorts for research projects, and so on. We have found that our various behavior descriptors, such as the Minnesota Multiphasic Personality Inventory (MMPI) and the Minnesota-Hartford Personality Assay (MHPA), which provide us with quantified scores on various aspects of behavior, such as anxiety, depression, antisocial behavior, can be utilized in a stepwise multiple discriminant function analysis to provide predictor models for these purposes. We have recently added in multiple regression weights for these same factors to provide us with level of pathology scores within the various discriminant function groupings.

In one of our current research projects, these techniques have been used to provide a matched comparison group of patients from the general hospital population for a specific group of experimental patients. These "twins" were matched on the basis of sex, age within 3 years, and self-description on admission to the hospital, utilizing the discriminant function classification of the MMPI profiles. In the particular research problem, we were trying to find a comparison group for the experimental group, since the particular experimental intervention made a blind or double-blind procedure quite impossible. Comparing course in hospital and final outcome for the twin pairs has given us, we feel, a reasonable basis for making claims about the success or failure of the experimental intervention.

A further refinement has been provided by the regression score for each individual

```
ALCOHOLICS     WEDNESDAY 5/8/74     3:30 PM

THE DIAGNOSIS SEARCHED FOR ARE:
291
309.13
301.81
303

PATIENT # AND NAME  BR 2   IGER, H.      18      01/29/74
(1)740129*304.4*304.0*301.77*R308.4   (2)740215*301.81*403.0*304.5  (3) (4)

PATIENT # AND NAME  FU 2   ROBBINS, J.  16      09/13/73
(1)730913*301.81*R301.77    (2)730927*308.4*301.77   (3)  (4)

PATIENT # AND NAME  BU 2   MOSKOVITZ, D.  20    03/29/74
(1)740329*303.1*304.5*R295.5   (2)740419*303.1*304.5*301.81     (3)  (4)

PATIENT # AND NAME  BU 1   MILLER, L.   49      11/16/73
(1)731116*309.8*303.2*301.4  (2)731227*303.2*301.4*R309.6*R309.3  (3)  (4)

PATIENT # AND NAME  DBS1   MARIKAR, M.  54      03/25/74    05/10/74
(1)740325*300.4*303.1*531*713  (2)740415*303.2*296.2*301.81*713   (3)  (4)

PATIENT # AND NAME  W.H.   KAPUCHINSKI, S.  40   01/14/74
(1)740114*300.4*303.2*301.81   (2)740201*301.81*300.4*303.1  (3)  (4)

PATIENT # AND NAME  COLE   MARIKAR, M.  71      05/18/73
(1)730518*303.2*R303.0*R303.1*300.4*R291.5   (2)730530*303.1*300.4*294.1 (3)(4)

PATIENT # AND NAME  TO 1   SILVERMAN, S.  51   01/25/74
(1)740125*303.2   (2)740214*303.2*300.4   (3)   (4)

PATIENT # AND NAME  COLE   GARNEAU, Y.  49      03/11/74
(1)740311*303.2*300.4   (2)740401*303.1*300.4*301.4   (3)    (4)

PATIENT # AND NAME  BRD2   BROWN, R.    47      12/18/73
(1)731218*296.2*R300.4*301.81*303.2   (2)740114*303.2*300.4*301.81*R354 (3)(4)
```

Figure 6. Output extractor program, alcoholics.

within the groups. This has allowed us to determine the initial level of pathology described by each patient and compare this with the final level of pathology, again in a nicely quantified numerical fashion that lends itself to statistical analysis far more satisfactorily than the three-step or five-step levels of improvement usually utilized in most psychiatric research.

In my opening remarks I indicated that cost effectiveness was one of the considerations to be discussed, as it has been throughout this conference. The cost to the hospital of the system that is now running, with all of the peripheral equipment when the full 40-terminal system is completed, will be approximately $650,000. All of this equipment has been purchased by the hospital and could be depreciated on a 5-year, or 10-year basis, more likely 10 years, so that the annual cost would be about $70,000 when equipment upgrading or replacement is added in. The cost per year for equipment maintenance and annual supplies is approximately $60,000.

Since we are still in the process of further development of the system, the personnel costs for the Data Processing Department are high, since additional systems and programming staff are required for continued development of new applications. The annual payroll figure of $224,000 includes all personnel employed in the data processing operation, both for business applications and clinical applications, in addition to development of activities. The approximate total cost of all the data processing activities, therefore, is $1,000 a day. Based on our average census of 385 patients, the cost per patient day is approximately $2.60. This is remarkably close to the goal we set some 8 years ago of a cost of $2.00 per patient-day for the data processing system, considering the inflationary pressures of the last 5 years.

Unfortunately, cost effectiveness is difficult to derive from the above figures, since we have not been in operation, even with a partial system, for a long enough period of time to measure the advantages from simultaneous distribution of information from its source to a number of necessary recipients in a single transaction, as against the old method of generating a number of individual messages. An example of this in the present system is the pharmacy transaction. From the single transaction in the pharmacy creating the label for the medication to be dispensed, the patient's medication record is updated, the charge entered goes to the patient's bill, and drug inventory is decremented by the amount of the medication dispensed, and various research files, for example, a file containing information on drug use in depressed patients, are updated. We are convinced, granted without hard evidence, that if the savings in clerical time and other logistics alone were taken into account, the net cost of the system would be minimal, and might even represent a saving.

In conclusion, we have tried to describe for you the computer system presently in use at the Institute of Living and to give you some idea of the issues that entered into our decision to develop the system. The scope of the data base, our plans for its future expansion, and examples of its utilization in different types of inquiry of the system have, we hope, provided useful examples of the flexibility, relative simplicity, and speed, in terms of response time, that a good computer-based patient information system can provide.

REFERENCES

1. Barnett, G. O. The modular-hospital information system. In: *Computers in Biomedical Research*, edited by R. W. Stacy and B. D. Waxman. New York: Academic. vol. IV. In press.
2. Greenes, R. A., A. N. Pappalardo, C. W. Marble and G. O. Barnett. Design and implementation of a clinical data management system. *Comput. Biomed. Res.* 2: 469–485, 1969.

Use of table file structures in a clinical research center

B. J. RANSIL [1]

Harvard Medical Unit, Boston City Hospital
Boston, Massachusetts 02118

PROPHET is an NIH-sponsored biological-chemical information system that incorporates a variety of computer science techniques—time sharing, extensible languages, data management, interactive graphics, graph-matching, molecular model building and statistics—into a user-oriented, user-responsive information handling tool to implement and facilitate drug research. Sponsored by the Chemical/Biological Information Handling Program of the NIH, Division of Research Resources, PROPHET is being designed and developed by the Medical Computer Systems Group of Bolt Beranek and Newman, Inc., Cambridge, Massachusetts (software) in collaboration with First Data Corporation, Waltham, Massachusetts (hardware) and a user group of diversified computer experience and research goals.

OVERVIEW

The system hardware consists of a time-share PDP-10 connected by dedicated telephone lines to terminals consisting of a Computek 400 display module, a Tektronix 4601 hardcopy unit and a Computek tablet/pen, a configuration with versatile, interactive input–output capability that has proven to be remarkably congenial to many drug research applications.

One unique PROPHET design requirement is that any investigator, with or without any previous computer experience, can be able to analyze his own research data directly and efficiently. Accordingly PROPHET is built around a set of *data types* including TABLE, ARRAY, GRAPH, MOLECULE, MULTI-MOLECULE, SAMPLE, NUMBER, TEXT, CHEMTEXT, COMMANDLIST, and PROCEDURE which are manipulated in simple English commands of specified syntax. While all data types may be stored by the SAVE command, TABLE is the only data type automatically and permanently stored on disk as a file structure until removed by a DELETE command. Using commands and data types, the user proceeds in stepwise fashion to tabulate, analyze, graph, curve-fit, display and print his data with no need for additional programming or knowledge of the internal structure of the data types being used. A listing of current PROPHET features and computing repertoire is provided in Table 1. Spelling errors are either machine corrected with completion of the command, or the command is aborted. Appropriate error messages point out errors in syntax, variable naming, shortages of desk space and the like.

[1] Present address of author: Dept. of Medicine, Beth Israel Hospital, Boston, Mass. 02215

The more computer-wise investigator may extend his analytical and data processing capability far beyond the current (ever-growing) repertoire by writing procedures in PL/PROPHET, a programming language syntactically modelled after PL/1.

TABLE 1. PROPHET features and computational repertoire

Command types

Display	Computations
Table	Text
Graph	Procedure
Molecule	General

Data types (variables)

Table	Number
Array	Text
Graph	Chemtext
Molecule	Commandlist
Multi-molecule	Procedure
Sample	

File structures (permaneltly stored on disk)
Table

Mathematical functions

ABS ....absolute value	SIN ...sine
SQRT ..square root	COS ...cosine
LOG10 .log base 10	SIND ..sine, degrees
LOG ...natural log	COSD .cosine, degrees
EXP ...exponent	MIN and MAX

Statistical commands
MAKE SAMPLE
TEST NORMAL
t-Test of means

Curve fitting
Fit line to curve (with plot) or table columns
Fit polynomial to curve (with plot) or table columns

Public procedures
3-Dimensional plotting
Simultaneous linear equation solving
Matrix and vector manipulation
% points—digitalization of data from curves
Multiple exponential fitting
Multivariate linear regression
Analysis of variance
Area under curve by trapezoidal rule

Programming
Intra-PROPHET with PL/PROPHET
Extra-PROPHET with FORTRAN

MAKING TABLES

Analysis of numerical research data proceeds by its storage into numbered, headed columns and numbered, named rows of named tables via the MAKE TABLE dialogue. The dialogue is initiated by typing the command: MAKE TABLE MAI ▲, where "MAI" is the table name and "▲" is the "Go" key which executes the command. Upon execution the screen returns the question: ARE THERE ROW NAMES? to which the operator responds Y(ES) or N(O). (Henceforth the "go" symbol will be omitted.) Upon execution the screen displays the prompt: COLUMN 1 HEADING = and the user types in an appropriate column heading for column 1. The next dialogue cue is COLUMN TYPE= to which the user responds with one of the mulitple choice responses appearing at the bottom of the screen: NUMBER TEXT CHEMTEXT DESCRIPTOR MOLECULE TABLE GRAPH. Upon execution the screen then displays the following choices: DERIVED ENTER NEXT EXIT RESTORE. DERIVED enables the user to construct the column from any previously entered column(s) expressed as arguments in algebraic relationships. ENTER cues the user through number-by-number data entry. NEXT moves the user on to the next column. EXIT stops the table-making process and displays the table. RESTORE stops table-making and restores any display that existed prior to table-making.

Corrections are made by typing a SET command: SET COL 1 ROW 2 OF MAI TO 63; or by typing SET, using the graph/pen and screen cursor to locate the offending table cell and typing the correct number.

Columns and rows of new data may be added indefinitely with the ADD COL and ADD ROWS commands that trigger off dialogues similar to the MAKE TABLE dialogue. Entire columns and rows or portions of them may be added to one table from another by appropriate commands. Any column or row may be deleted by the DELETE command: DEL COL 3 OF MAI, or DEL ROW 6 OF MAI.

MAKING GRAPHS FROM TABLES

Plots may be made from table columns using either the MAKE GRAPH dialogue or the command MAKE GRAPH FROM MAI AS COL 1 VS COL 7. The MAKE GRAPH dialogue prompts the user step by step to specify the title of the graph, the *x* and *y* axis labels, the table it is to be constructed from, the column to be used for the *x* axis values, the columns to be used as *y* values and appropriate symbols and labels.

Supporting each graph are three so-called systems tables controlling all information relating to *1*) axes, *2*) curves and *3*) points of the graph. One may change any aspect of the graph (labels, symbols, connectors, line width, tick placement, number of ticks per inch, origin placement, highest and lowest values for *x* and *y* axes, log or linear scale, right *y*-axis, etc.) simply by making the appropriate change in the systems tables using a SET command.

Curves from one graph may be added to another by the command ADD CURVES 1,3 TO GMAI FROM GJAL. Curves may be deleted by the command DEL CURVE 1,2,6 FROM GMAI.

Graphs are readily displayed on execution of the EXIT command from the graph dialogue. The size of the graph may be adjusted at will with the SCALE command and an appropriate scale factor, e.g., the commands SCALE 1.5 and SCALE .65 enlarge the graph 50%, and decrease it 35% respectively. One may scale separately for the *x* or *y* axis, or any portion of them. Once displayed, copies of graphs and tables or any other material on the screen may be copied by pushing the "copy" button on the hard copy unit or by typing the PRINT command which displays the date as well.

These brief descriptions by no means exhaust the table-making, graph-making and display capabilities of PROPHET, but do provide some idea of its operational simplicity and flexibility. One major feature, MAKE MOLECULE, is outside the scope of the topic and is therefore omitted entirely.

DATA ANALYSIS

While PROPHET can perform perfunctory calculations more aptly relegated to desk and hand calculators, its greatest computational power at the command language level resides in the ability to perform mathematical and statistical operations from rows and columns of tables. With the derived column feature one may perform a variety of numerical analysis techniques based on tables, such as data linearization, numerical integration and differentation and table interpolation. Addressing specific table columns, least squares straight line and polynomial fits to data are readily obtained with the respective commands: FIT LINE TO COL 1 VS COL 2 OF MAI (which supplies full regression and goodness-of-fit data); and FIT POLYNOMIAL TO MAI (which triggers off a dialogue asking for lowest degree to be fitted and for the number of fits). An exponential fit is performed using a "public procedure" (a procedure available to all users via a shared "public" space) entailing 4 commands: DECLARE GMAI GRAPH, RESTORE PUBLIC %EXPFIT, COMPILE %EXPFIT, CALL %EXPFIT (MAI, GMAI). A short dialogue then ensues wherein the user specifies the number of exponents to be fitted and the range of points desired to compute initial estimates of the curve fitting parameters. Output consists of a graph of the curve fit, and tables of experimental data, calculated data, and goodness-of-fit data.

DATA MANAGEMENT

Occasionally a research project requires the construction and manipulation of a data bank that is indefinitely extensible in two or more variables. This is achieved in PROPHET for two variables by indefinite extension of table columns and rows. Additional variable capability is achieved by the "storing tables in tables" feature wherein each cell (possessing a specific row–column index) may hold a table with a subordinate row–column index system equivalent to a third and fourth variable. Data management reduction and manipulation of large tables and tables within tables can be tedious and inefficient for large data accumulations and large numbers of variables, but this is readily circumvented by constructing accounting and processing procedures based on index numbering systems, sorting procedures and commandlists using PL/PROPHET. Programming assistance is available from Bolt Beranek and Newman staff members upon request.

PUTTING PROPHET TO WORK

A typical research project involves the clearance (disappearance) of radioactively labeled insulin from the plasma of four patients groups corresponding to *1*) a healthy control group, *2*) a group of patients with little or no renal function, on dialysis awaiting transplant, *3*) the same group immediately after removal of both kidneys (nephrectomy), and *4*) the same group 6 months after a successful transplant and restoration of renal function.

The experimental data consisted of plasma insulin concentrations at specific time intervals over a 3–4 hour time span for each patient. These were entered into four PROPHET tables (named *A*,*B*,*C*,*D*) corresponding to the four subject

{ }PRINT{ DATE: 5/23/74 }

MNSINCLEAR 15R X 11C

	1 TIME (MIN)	2 CONTROL % CLEAR	3 PRE-NEPH % CLEAR	4 POST-NEPH % CLEAR	5 POST-TRN % CLEAR
1	0.	100.	100.	100.	100.
2	2.	98.55	98.24	97.56	95.48
3	5.	97.75	96.19	95.84	92.7
4	10.	90.7	90.7	89.21	85.24
5	20.	78.17	83.27	84.4	76.78
6	30.	74.7	80.58	80.02	73.88
7	40.	68.95	78.16	76.71	71.41
8	50.	65.7	76.67	75.81	70.31
9	60.	64.72	76.5	75.78	69.17
10	90.	61.69	73.12	74.61	65.1
11	120.	58.66	71.75	72.02	61.4
12	150.	56.44	70.21	71.7	60.92
13	180.	54.84	68.35	69.48	61.79
14	210.	58.05	68.17		55.72
15	240.	51.17	67.5		

Figure 1. Mean insulin clearance values.

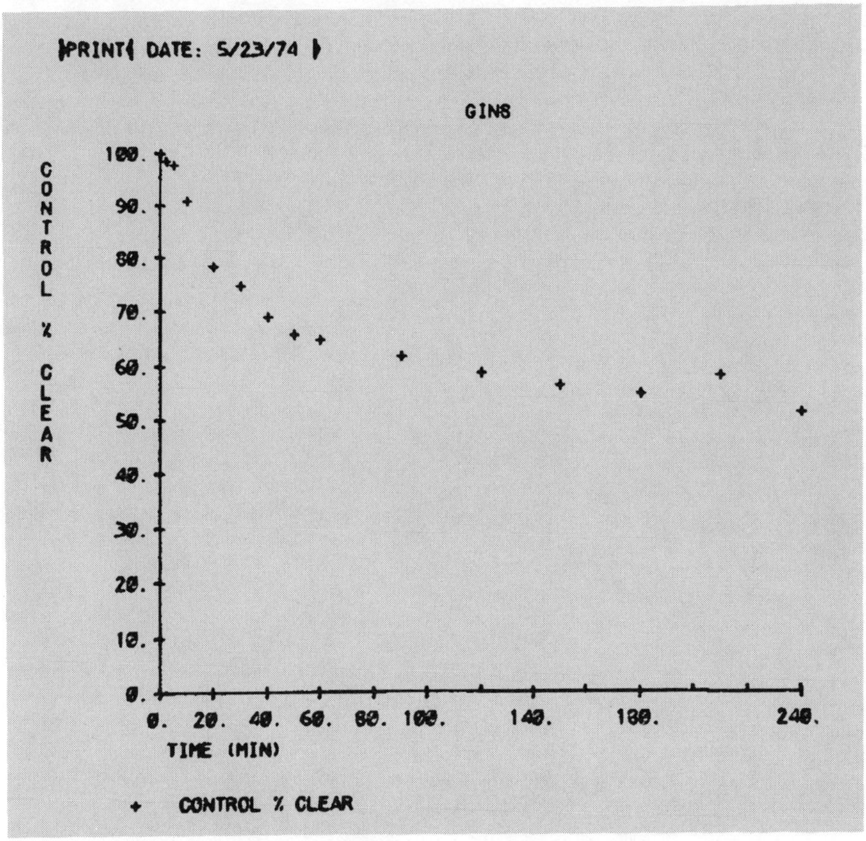

Figure 2. Insulin clearance values in control subjects.

groupings, in which column headings were TIME and the relevant patient names. Column means were then computed with a procedure that calculates column and row central measures for any table, adding them to the original table. The desired means now exist as *column 6* in each of the four tables. A master table of means (called MNSINCEAR) is now constructed from *A*,*B*,*C* and *D* with the command MAKE TABLE MNSINCLEAR FROM A FOR COL = 1,6, (where *Col. 1* contains time values and *Col. 6* contains means) and three successive commands ADD COL TO MNSINCLEAR FROM *B* (or *C*, or *D*) FOR COL = 6. The result is MNSINCLEAR comprising 15 rows and 5 columns containing time and mean insulin clearance values (Fig. 1.)

The investigator next visualizes a typical disappearance curve (to decide upon an appropriate curve-fit function), with the command MAKE GRAPH GINS FROM MNSINCLEAR AS COL 1 VS COL 2. The command DISPLAY GINS yielded an exponential curve (Fig. 2) requiring two, perhaps three exponents. Accordingly the %EXPFIT procedure is called, compiled and applied to the 4 curves, using double exponential fits, yielding a graph, goodness-of-fit data and a table of values calculated from the best-fit function. The latter shows a significant over-shoot at t = 0 for 3 cases, so a triple exponential fit is computed for all 4 cases. Comparison of goodness-of-fit data for the double and triple exponent solutions suggests that the triple-exponential fit is a better fit for the 3 groups with kidneys present whereas the double-exponential fit is the better fit for the group without kidneys.

The analysis tells the investigator *1*) kidneys appear to play a role in insulin clearance; *2*) insulin clearance with kidneys present is compatible with a triple-compartment (or triple first-order process) model; and *3*) insulin clearance without kidneys is compatible with a two-compartment (or double first-order process) model. The actual role two healthy kidneys play in insulin clearance is extracted from the data by subtracting the calculated nephrectomized data from the calculated control data, using the ADD

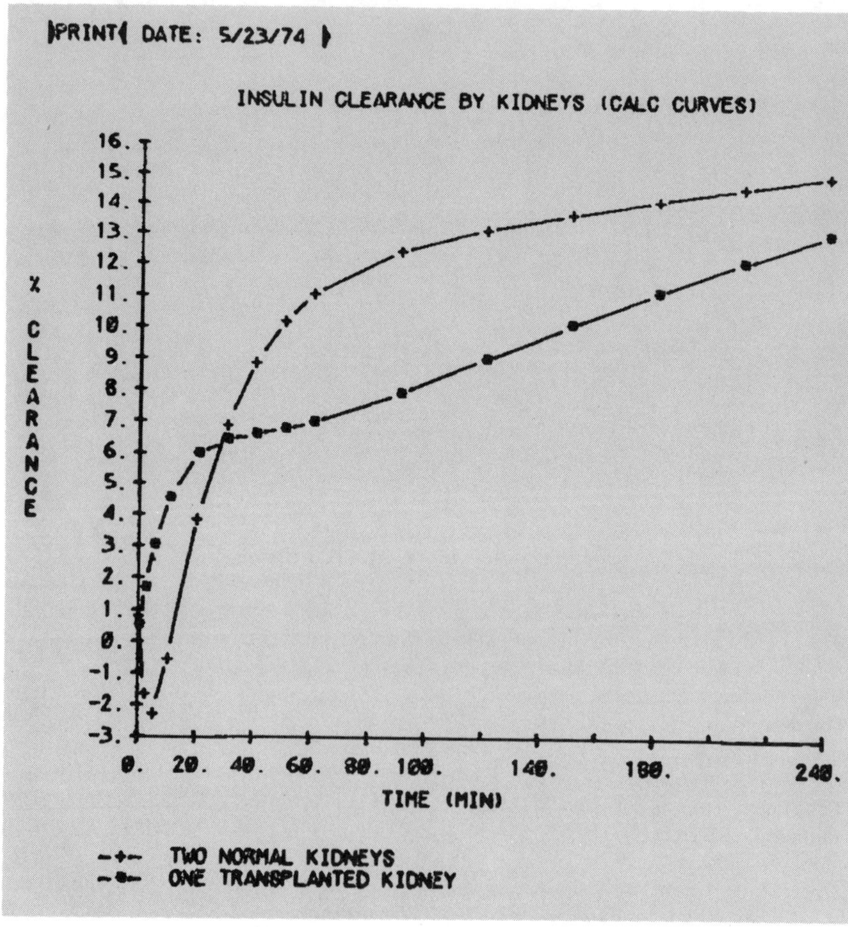

Figure 3. Renal component of insulin clearance.

COLUMN command and the DERIVED option. The role one functioning transplanted kidney plays in insulin clearance is obtained in a similar manner. The final results are quickly graphed and displayed, yielding Fig. 3.

Total analysis time amounted to approximately 1.5 to 2.0 connect hours and about 10 CPU min, with average response time. The interactive nature of the process—between graphs and goodness-of-fit data—enabled the investigator to comprehend his data rapidly, make decisions quickly and carry the analysis to a quick, satisfactory conclusion, with data permanently stored for checking and additional analysis following digestion and study.

COST-BENEFIT FACTORS

From the user's point of view a cost-benefit analysis of PROPHET performance in the laboratory depends on the identification and measurement of a number of variables for which at best only anecdotal data are available. One may utilize such crude efficiency estimates as research output and operational costs. For example, one group at this site reports an increase in output due to PROPHET of 33%—with a corresponding increase in materials cost! Yet other real benefits exist that defy measurement, such as better utilization of research data, and enhancement of the scientific outcome of a research project. Because it can provide "instant answers" to almost any conventional biometric test, PROPHET encourages the investigator to ask more searching questions of the data, and to look at it from perspectives he ordinarily would not explore if he had to crank through the number-crunching steps himself.

CONCLUSION

In summary, the table-structured PROPHET, still in process of development, has proved itself to be a truly user-oriented, research-stimulating tool in this medical research environment primarily because it brings computer power to the finger tips of the non-computer trained investigator, allowing him to interact with, and analyze his data in a manner never before possible.

REFERENCES

1. Castleman, P. A., C. H. Russell, F. N. Webb, C. A. Hollister, J. R. Siegel, S. R. Zdonik and D. M. Fram. Implementation of the PROPHET system. *Proc. Natl. Comput. Conf.* 43: 457–468, 1974.
2. Hollister, C. A. *Users Manual for the PROPHET System.* Bolt Beranek and Newman, Inc., periodically updated.
3. Johnson, C. L. Applications of PROPHET system in molecular pharmacology—structure-activity relationships in monoamine oxidase inhibitors. *Proc. Natl. Comput. Conf.* 43: 473–476, 1974.
4. *Public Procedures Notebook.* Technical Document No. 47, Bolt Beranek and Newman, Inc.
5. Ransil, B. J. Applications of PROPHET in human clinical investigation. *Proc. Natl. Comput. Conf.* 43: 477–483, 1974.
6. Weeks, C. M., V. C. Cody, S. Pokrywiecki, D. C. Rohrer and W. L. Doux. Applications of the PROPHET system in correlating crystallographic structural data with biological information. *Proc. Natl. Comput. Conf.* 43: 469–472, 1974.

Session IV—Data base systems
Rapporteur's summary

DELPHIS C. GOLDBERG
Intergovernmental Relations Subcommittee
Washington, D. C. 20515

The session on Data Base Systems covered a variety of computer applications in biomedical research and clinical medicine. These relate to three large systems (the 3M Company, Chemical Abstracts Service[1], and Permanente Medical Group), and two relatively small systems (Institute for Living and Boston City Hospital). One of the large and two of the small systems are directly related to optimizing the health care of patient populations. In the case of the Boston City Hospital, which participates in the NIH-funded PROPHET system, research is a primary objective.

While the motivating force for developing data systems may differ among these organizations, there appears to be the common objective of increasing the productivity of their investigators or clients by making adequate data available in usable form. This is seen most readily in the case of the 3M Company and Chemical Abstracts Service, both of which are confronted with the pressure of dealing economically and expeditiously with scientific literature that is growing at an explosive rate. Comprehensive literature searches are essential, as one panelist expressed it, to keep from reinventing the wheel every 5 years. The 3M Company has the added incentive, as a large industrial firm, to keep competitive by acquiring current information as completely and rapidly as possible in areas such as new chemical structures and patents issued throughout the world.

It seems clear from the presentations of the panelists that large data base systems can be developed and managed successfully in a corporate setting where there is a strong interest in controlling costs and increasing productivity, and where the nature of the data makes it relatively easy to compare costs under alternative operating conditions. Obviously the situation is more complicated when one evaluates cost-effectiveness in a clinical and/or research setting where the variables are more numerous and organization is less structured.

I think it is important, at this point, to distinguish between the research and development phase and normal operating activities in judging the cost-effectiveness of a computer facility. It may be in the national interest, for example, for the Federal Government to underwrite the cost of developing data base systems

[1] Papers entitled *Use of computers for technical communication in support of a large research staff*, by Julianne H. Prager of the 3M Company, St. Paul, MN, and *Management of a large data base of chemical literature information*, by Gary A. Wilson of the Chemical Abstracts Service, Columbus, OH, were presented at the Computer Conference, but were not submitted for publication.

designed to improve the delivery of health care. In subsidizing projects of this kind the public has the right to expect that the projects selected will be suitable for replication in a substantial proportion of the health care institutions of a given type.

Once the R & D phase is completed and the facility has been shown to be cost-effective, it would seem reasonable to expect the hospitals or other users to recover their data facility costs through service charges. The application of a developed system in other institutions might require an initial government subsidy for construction of the data base to the point of operational readiness.

Examples were presented of the contribution of data base systems to improved patient care. In the case of the Permanente Medical Group, it was stated that the multiphasic screening program has increased physician productivity by 50% while benefiting patients through early detection and early treatment of medical conditions.

The data base system developed at the Institute for Living, although incomplete, is believed to have improved the effectiveness of patient care. It was acknowledged, however, that good baseline information is not presently available on the appropriate length of a hospital stay for various conditions. Such information, it was suggested, will have to be developed to satisfy the requirements of the PSRO program and third-party payment arrangements.

In a related area, not specifically covered by the panel, the importance of retrievable patient records is becoming increasingly evident as new medical knowledge makes it desirable to locate individuals previously exposed to chemical substances or medical procedures that are found to be extremely harmful. A notable example along these lines was the excellent epidemiological work done several years ago at the Massachusetts General Hospital in establishing a link between the ingestion of diethylstilbestrol by women treated for threatened abortion in the 1940's and 1950's, and an unexpectedly high incidence of vaginal cancer in their daughters 20 or so years later. While the urgency of a nationwide effort to identify DES-treated patients and their daughters is obvious, adequate medical records simply do not exist except in a limited number of institutions and a few states. Computerizing medical records, it should be noted, also involves the sensitive issue of privacy.

In the hierarchy of computer facility users, the individual investigators as a class are probably the least susceptible to institutional pressures for accountability and productivity. Here the problem appears to be different and perhaps more difficult. Fortunately, for me, this was not a matter within the purview of the panel and I shall not attempt to discuss it.

Session V

Networks and shared facilities

Introduction

PETER G. LYKOS, *Organizer and Chairperson*
Illinois Institute of Technology
Chicago, Illinois 60616

If the term "computer networking" were to be interpreted literally one would conjure up an image of a number of computers interconnected such that any one computer is not only interfaced to one or more other computers but that functionally one could envision a single problem being worked on by two or more computers in the net with no human intervention.

Actually so-called computer networks commonly found today are of two general types: *1*) star computer networks whereby a single central computer has a number of communication lines radiating out to some sort of terminal devices; and *2*) communication networks, that is, using the telephone system whereby a user at a terminal may dial any one of a large number of computer-based services remotely located.

From a technical point of view, all of the components of computer networking could be present in a single laboratory as well as spread over a campus, a region, the nation, or over the entire world.

Some of the motivations for computer networking on a regional or national scale have already been mentioned in the earlier sessions of this Conference. The most immediate and least politically sensitive is the sharing of data and bibliographic information resources. In addition there have been mentioned specialized resources, dedicated to some particular mode of problem solving, usually discipline oriented, that are expensive and whose users are geographically distributed. A third motivation is to provide support for the experimentalist, i.e., hierarchical computing and laboratory automation. A more subtle motivation is a desire for the standardization of protocols and algorithms where the variations that now exist that render software from different sources incompatible stem from idiosyncratic differences among researchers rather than from substantive differences. Finally an important motivation, also already suggested earlier in the Conference, is that computer programs represent implementation of real-world problem-solving procedures. Sharing of such algorithms via a small number of sites creates a situation where an accumulation of the best thinking of the researchers working in the specific area cannot only be accessed and used by others within a common denominator or mode of representation but also be susceptible to continual revision with input from all concerned. Computer-based conferencing is an obvious offshoot.

A large number of computer network related activities are already underway—too numerous to mention here. However, two related events should be brought to the attention of this group. The first is the EDUCOM Conference of 1966, which resulted in the concept of EDUNET—a computer communication complex spanning the country and supportive to education and research. Leaders in that initiative were from the life sciences. EDUNET never happened. Recently, in July, 1971, the EDUNET concept was rejuvenated. There was begun at NSF an NSF initiative called A Feasibility Study of a National Science Computer Network. One of the components of that initiative was a set of three seminars organized and conducted by EDUCOM. A preliminary report appeared in *Science*, October 5, 1973, and the complete report, published in 1974 by MIT Press, is entitled *Networks for Research and Education*. The principal conclusions reached in those broadly based and intensive seminars were as follows: *1*) computer networking must be acknowledged (no person is an island); *2*) major inhibitors of realization of networking potential are political, organizational and economic—not technological; and *3*) networking by itself is not a solution to current deficiencies—user practices, institutional procedures and state and federal government policies need to adapt.

Several recommendations came out of those EDUCOM seminars, including the formation of a Planning Council to formulate a specific business plan and to proceed to implement a facilitating network. That Planning Council has been formed and is beginning its work.

It is my feeling that demonstrated usefulness and research enhancements—discipline by discipline—of computer network augmentation of research is the essential precursor to the construction of a super national system such as is outlined in the EDUCOM report. That initiative has to come from the disciplines themselves and only now are we beginning to see evidence of their interest.

In an attempt to give you a firsthand account of several important aspects of computer networking augmentation of research within the constraints of one half-day session at this conference, we have five complementary papers supplemented with a brief synopsis of a sixth.

First, Dr. William Raub of NIH and the PROPHET system illustrates a star network initiative by a federal agency supporting extramural research in the life sciences.

Second, Professor Rodney Harrington of Purdue illustrates how a computer user got into a minicomputer mininetwork situation supporting multifaceted laboratory research.

Third, we have Professor Edward Bowdon, Sr., an observer and analyzer of computer networks from an economics point of view who got started in this line of endeavor before it became popular to do so.

Fourth, Frank Heart, an electrical engineer by training, role of the computer in medical research by vocation, now a computer sciences administrator in a commerical organization, Bolt, Beranek and Newman, Inc., which is the principal contractor for the ARPANET. Frank is on the board of directors of TELENET, a new wholly-owned subsidiary of BBN which is creating a commercial data network based on ARPANET technology.

Fifth, Professor Ed Meyer will describe a National-Laboratory-based effort, CRYSNET, a crystallographic computing network with interactive graphics display, which represents a cooperative effort between researchers in different administrative settings: a National Laboratory, a nonprofit Research Institution, and a University.

Finally, Professor Ashenhurst will give us a brief synopsis of his work on hierarchical computing and laboratory automation. A distinguishing feature of this work is that the initiative and perspective are from a computer scientist.

The PROPHET System and resource sharing

WILLIAM F. RAUB
National Institutes of Health
Bethesda, Maryland 20014

The PROPHET System is a specialized computer resource sponsored by the National Institutes of Health (NIH). The short-term objective for the System is that it be a unique and valuable research tool shared by a broad array of geographically distributed scientists studying chemical/biological interactions. The long-term goal for the System is that it promote the emergence of a predictive science regarding drugs and drug-related phenomena. This paper discusses this shared research resource and the management considerations associated with its operation and continuing development.

BACKGROUND

Understanding chemical/biological interrelationships is central to attacking many major human health problems. Whether the objective is arresting tumor growth, stabilizing blood pressure, preventing conception, or neutralizing the effects of environmental poisons (to name only a few), significant progress is unlikely to occur until more is known about how chemical substances and life processes influence one another. A knowledge base sufficient to predict the consequences of specific chemical/biological interactions is the grail toward which pharmacology, toxicology, medicinal chemistry, biochemistry, and a host of related disciplines are oriented.

The route to new knowledge about chemical/biological interactions, of course, is research; and relevant investigations now occupy the energies of thousands of scientists throughout the world. These studies collectively span the full range of biomedicine (i.e., from molecular biology to patient care) and employ not only animal and human subjects but also a rich array of supporting technologies. Hardly any area of human inquiry is more protean in character than the interface between chemistry and biology.

One prominent characteristic of most studies of chemical/biological interactions is their highly empirical nature. Data generally are heterogeneous and in plentiful supply. Well-developed conceptual models usually are found only in tightly circumscribed domains. Investigators frequently are able to amass observations more easily than they can account for them in terms of any particular theory or formalism. Representation, organization, manipulation, and analysis of data not only occupy a significant proportion of research time but often are the rate-limiting steps.

Computer science and technology offer increasingly powerful tools both for dealing with these information-handling problems and for promoting the emergence of better predictive capabilities. Experience in this and other areas of natural phenomena has shown that data arrayed in structured form(s) generally are more useful than data which are unstructured, and that structured data which also are machine manipulable almost invariably present more opportunities for hypothesis generation and testing than is the case with exclusively manual methods. Similarly, when working hypotheses are expressed and exercised as procedures or computer models, one frequently can expect to see accelerated development of a theoretical base for the science involved. The domain of chemical/biological interactions is an especially appropriate arena in which to exploit the logistical and cognitive assistance that computers can supply to a prepared mind.

In recognition of the above, the NIH launched in 1967 its Chemical/Biological Information-Handling (CBIH) Program. Since that time, through a series of carefully programmed efforts, NIH staff and advisors have sought to stimulate interest in and use of computer methods for the study of chemical/biological interrelationships. This effort has required a blending of the strengths and skills of academe, the private sector, and Government. A major part of the task has been finding ways to instill appreciation of computer-based techniques in scientists whose training and experience does not include significant exposure to quantitative methods in general and mechanized symbol manipulation in particular.

The principal achievement of this NIH program is the PROPHET System. Through the operation and continuing development of PROPHET, NIH offers to biomedical scientists throughout the nation (even those who are computer naive) a unique opportunity to evaluate computer-based information-handling methods in their search for chemical/biological interrelationships. Moreover, by successfully accommodating an ever-growing community of scientists distributed widely in geographic location but united intellectually through common and complementary interests, PROPHET is emerging as a paradigm for the kinds of shared national research resources that in all likelihood will be a hallmark of the science of the future.

DESCRIPTION OF THE PROPHET SYSTEM

PROPHET is a time-shared computer system accessible remotely via graphic display terminals communicating over dedicated, voice-grade telephone lines (Fig. 1). Users exercise PROPHET's capabilities through English-like commands oriented to the nonprogrammer scientist and through procedural state-

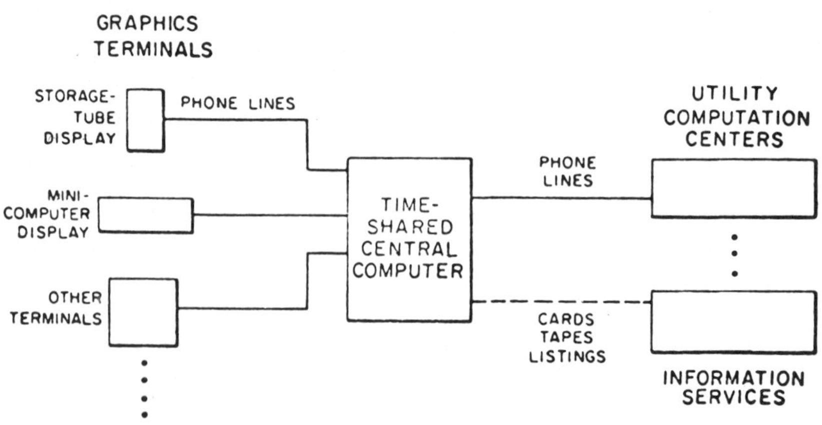

Figure 1. PROPHET system equipment organization.

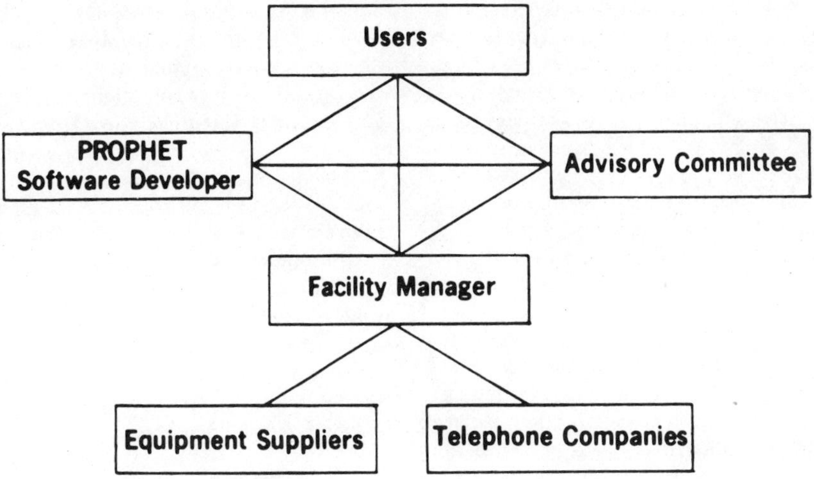

Figure 2. Relationships among participants in the PROPHET project.

ments closely akin to PL/I; these commands and statements can be intermixed freely to handle specialized applications. PROPHET features data types that represent such common constructs as data table, graph, molecule, and statistical sample. Software for management, display, and analysis of data has been constructed so as to allow users to invoke complex operations through simple commands. An extensible base language with facilities for defining new data types and new command syntax simplifies development. More detailed information on these various aspects of the PROPHET System is provided elsewhere (1–3, 5, 7, 8).

TABLE 1. PROPHET users as of June 1974

1. Department of Pharmacology
   School of Medicine
   University of Pittsburgh
   Pittsburgh, Pennsylvania

2. Harvard Medical Unit
   Thorndike Memorial Laboratory
   Boston City Hospital
   Boston, Massachusetts

3. Department of Pharmacology
   Mount Sinai School of Medicine
   City University of New York
   New York, N.Y.

4. Molecular Biophysics Department
   Medical Foundation of Buffalo
   Buffalo, New York

5. Research Institute on Alcoholism
   New York State Department of Mental Hygiene
   and
   State University of New York at Buffalo
   Buffalo, New York

THE PROPHET USERS

The PROPHET users collectively have interests ranging from quantum pharmacology to human clinical investigation, and they exercise thoroughly almost every facet of the System. Both the number of user sites and the diversity of user applications are expected to grow steadily for the foreseeable future. The locations of the principal PROPHET users as of June 1974 are shown in Table 1. Titles of some selected user projects are shown in Table 2. More detailed accounts of applications also are available (4, 6, 9).

PROPHET users are selected in a highly formal way. Research prospectuses submitted by interested scientists in response to NIH announcements are appraised and rated by peer review panels according to published criteria. NIH staff then identify among the highest ranked prospectuses those that seem most closely attuned to the goals and needs of the PROPHET project and, within the constraints of hardware and contractor personnel capacity, arrange for provision of the requisite remote terminal equipment and telecommunications services. This procedure does much to insure that user needs and System capabilities are well matched and that PROPHET services are allocated strictly on the bases of scientific merit and CBIH Program relevance. The procedure also serves to identify whether there is a protagonist for PROPHET at the candidate user site—i.e., a highly motivated individual who can be expected to become expert in the use of the System, help NIH contractor personnel overcome problems with equipment and services at the site, and teach colleagues to appreciate System features that are not intuitively obvious to the uninitiated. The presence of such a protagonist almost invariably is the key to a highly successful installation.

RELATIONSHIPS AMONG PARTICIPANTS IN THE PROPHET PROJECT

The organization of the various performers in the PROPHET project is shown in Fig. 2. Continuing System design and

TABLE 2. Selected applications of the PROPHET System

University of Pittsburgh:
- Hallucinogenic drug action on somatosensory systems (primates)
- A phenotypic resistance to tetracycline (bacteria)
- Nervous control of urinary bladder and large bowel (cats)

Boston City Hospital (Harvard Medical Unit):
- Factors inducing the abnormal metabolism of trauma
- New concepts in parenteral protein-sparing therapy
- Epidemiologic studies and chronic respiratory diseases
- Eosinophil functions
- Clinical epidemiology of heart and lung disease
- Pathophysiology of the circulation
- Water and electrolytes in health and disease
- Computer feedback study of cardiac muscle mechanics
- Metabolism training program
- Nutrition training program
- Diseases of the liver and portal circulation

Mount Sinai School of Medicine:
- Psychopharmacologic activity—quantum chemical study (rats)
- Psychopharmacological changes following brain damage

Medical Foundation of Buffalo:
- Training program in endocrinology
- Molecular structures of thyroactive compounds
- Molecular structures of steroids
- Cancer of the thyroid and its hormonal identity
- Ionophores, transport and membrane structure
- Human development and estrogen biosynthesis
- Ultrastructure of antihypertensive prostaglandins

Research Institute on Alcoholism,
(New York State Department of Mental Hygiene):
- Structure/activity relationships of alcohol congeners: effects and withdrawal treatment
- Neuronal sites of ethanol action
- Tolerance to and dependence on ethanol in goldfish

TABLE 3. Tasks of the PROPHET software developer

Task	Full-time equivalents
Facilitating access to PROPHET	2.6
Improving PROPHET efficiency	1.7
Adding new PROPHET features	2.5
Documenting PROPHET	0.6
Management and other	2.4
Total	9.8

development arises from the interactions of these groups. Note that the software development and facility management functions are organizationally separate; this helps to achieve a viable balance between these equally critical but substantively distinct tasks. Note also that the facility manager is the "final common path" to and from the myriad suppliers of hardware and services.

The NIH staff have the responsibility for maintaining and modifying these consortium arrangements and for facilitating interactions among the performer elements. PROPHET users are selected through the process described above. The Advisory Committee is an officially constituted panel of NIH consultants (almost exclusively non-Federal) having expertise relevant to various aspects of the CBIH Program. The software development and facility management functions are effected via contractual agreements between NIH and profit-making organizations; these contracts are subjected to national recompetition periodically to insure as far as possible that the best available talent is being engaged. Equipment and services are acquired either directly by NIH or through the Facility Manager.

COSTS

As of June 1974, the PROPHET System has been in service for just over 2 years; inauguration of the service was preceded in turn by almost 4 years of design and initial implementation activity. The cumulative expenditure by NIH over this 6-year period was $4,881,778. Of this amount, $1,466,334 was used for the purchase of the central computer and maintenance services, remote terminals, and other hardware. The bulk of the remainder was used for direct labor and associated costs.

Another perspective on costs is gained by examining the ongoing expenses for System operation and continuing development. Tables 3 and 4 summarize the responsibilities of the PROPHET Software Developer and the Facility Manager, respectively, in the form of a breakdown of their tasks and the corresponding number of full-time personnel equivalents involved. Multiplying 17.4, the total annual personnel requirement, by $48,000 per man-year (an estimator which includes travel, overhead, general and administrative expense, and fixed fee) gives $835,000. Adding to this approximately $65,000 per year for hardware maintenance services and $40,000 per year for telecommunications charges makes the current annual level $940,000.

To date, the CBIH Program has borne almost all the costs of the PROPHET project. During Fiscal Year 1975 the PROPHET users are to begin sharing in the costs of System operation and maintenance. Nevertheless, significant subsidy by NIH will be required for the foresee-

TABLE 4. Tasks of the PROPHET facility manager

Task	Full-time equivalents
Operating the facility; coordinating hardware maintenance	3.6
Maintaining remote terminals; coordinating telecommunication services	2.2
Developing and maintaining PDP-10 system software	1.2
Management and other	0.6
Total	7.6

able future, especially with regard to further System development.

PROBLEMS AND PROMISES

There are several major areas of difficulty in maintaining the PROPHET System as a successful shared research resource. First, response time to user commands is often slower than desirable; the striving for "friendliness" (which manifests itself as very large programs) and continuing System growth (which requires that service and development work share the same hardware) places severe demands on the facility. Second, remotely located graphic display terminals and associated telephone lines rarely work satisfactorily without regular attention by highly trained contractor personnel; computer-naive biomedical scientists usually are unwilling to assume such a burden. Third and finally, the System generally requires a protagonist at a new site before it successfully penetrates the research milieu; the burden of proof remains with those who advocate computer tools for research into chemical/biological interrelationships.

Weighing both accomplishments and problems in the balance, the outlook for the PROPHET System seems encouraging. The stage is set for steady maturation as a nationally shared computer resource. There are no problems that shouldn't give way to good taste, careful management, and enlightened tenacity. There are more high quality, enthusiastic applicants than capacity to accommodate them. The basic PROPHET tools have stabilized to a considerable degree. The capability for graceful accretion of important information-handling functions developed outside the PROPHET project has been demonstrated. The PROPHET users are finding more and more areas where the System enhances their research productivity. And PROPHET-mediated sharing of programs and data bases among geographically separate groups is beginning to be practiced.

REFERENCES

1. ANNOUNCEMENT. The PROPHET System. *Federation Proc.* 32: 1744, 1973.
2. CASTLEMAN, P. A., C. H. RUSSELL, F. N. WEBB, C. A. HOLLISTER, J. R. SIEGEL, S. R. ZDONIK AND D. M. FRAM. The implementation of the PROPHET System. *Am. Fed. Inform. Proc. Soc. (AFIPS) Conf. Proc.* 43: 457, 1974.
3. HOLLISTER, C. A. *PROPHET System Users' Manual.* Bolt Beranek and Newman Inc.
4. JOHNSON, C. L. Applications of the PROPHET System in molecular pharmacology—structure/activity relationships in monoamine oxidase inhibitors. *AFIPS Conf. Proc.* 43: 473, 1974.
5. *PARSEC Users' Manual.* Bolt Beranek and Newman Inc.
6. RANSIL, B. J. Applications of the PROPHET System in human clinical investigation. *AFIPS Conf. Proc.* 43: 477, 1974.
7. RAUB, W. F. The role of specialized data structures in computer-based management, analysis, and communication of pharmacological information. Fourth International Biophysics Conference, Moscow, USSR, 1972.
8. RAUB, W. F. Fiscal Year 1974 Annual Report, Chemical/Biological Information-Handling Program, National Institutes of Health.
9. WEEKS, C. M., V. CODY, S. POKRYWIECKI, D. C. ROHRER AND W. L. DUAX. Applications of the PROPHET System in correlating crystallographic structural data with biological information. *AFIPS Conf. Proc.* 43: 469, 1974.

MIRACLE
a hierarchical data acquisition system for laboratory automation[1]

RODNEY B. HARRINGTON AND RONALD L. GIESE
Purdue University
West Lafayette, Indiana 47907

New advances in laboratory instrumentation have given scientists a new dimension to their research, but analysis and interpretation of the data generated by these instruments has increased in difficulty to the point where it has often become a limiting factor in their effective use. It was logical to turn to computer processing of these data to relieve the scientist of this burdensome task. Often the initial approach has been to consider the use of existing digital batch process computing. However, the turnaround problems of overburdened centers can be inhibiting; but even discounting this feature, analog-type data (unless analog processing is employed) must be transformed to digital information. Frequently this is in the form of punched cards. This approach fails to take advantage of interfacing the instrument directly to the computer and thus precludes on-line data acquisition, interactive control, or real-time data processing. Because of the batch and digital orientation of most large computing centers, the probability for linking instruments on-line with special interfacing and real-time demands at these centers, is low indeed, although it has been successfully accomplished at the University of Alabama under Dr. Josiah Macy and at the University of Utah under Dr. Homer Warner among other institutions.

Another approach (Iracki and Malmstadt (5) and Epler (4)) is the use of the small dedicated computer interfaced to a single instrument which is task-limited to that particular instrument or perhaps a small number of intralaboratory instruments. Since these computers operate independently and may be of different manufacture, it becomes difficult to upgrade the total system. When several of these small dedicated systems are installed in a larger research organization and economies examined, it is often discovered that greater capability and flexibility per unit investment may be achieved by another approach.

A third alternative utilizes a larger computer to serve several instruments. Downer (3) has described one example of such a system. This approach usually employs a type of real-time operating system with its attendant priority interrupt structure. Although these systems generally have more computing power and flexibility than the smaller dedicated systems, instruments may not be added or deleted easily from system to system.

[1] Approved for publication by the Director of the Purdue Agricultural Experiment Station as Journal Paper Number *5516*.

Yet another alternative is a hierarchical computer network. In a three-level system described by Klopfenstein (6) a series of minicomputers dedicated to a single instrument make up the first level. These computers are interfaced to a second level computer which acts as a communications link to a third level large computer. The large computer provides the computational power required for large scale calculations and large volume data storage. Ashenhurst (1) is developing a Minicomputer Interfacing Support System (MISS). The MISS facility is based on a hardware/software subsystem, the Minicomputer Operation Monitor, which furnishes direct on-line support to a battery of minicomputers. It also supports a set of peripheral devices, input/output and some mass storage. It is proposed that this system be further interfaced to a large scale computer network. K. Wilson at the University of California, San Diego (personal communication) is also developing a three-level hierarchy computer network. The first level of this system consists of a series of dedicated minicomputers. The second level of a combination of computers is planned which will provide disk storage capability for the first level computers plus a greater computational power and also a peripheral handler. This second level is interfaced as a user node in the ARPANET (7) system. These projects greatly enhance the level of computing power available to the laboratory worker by providing a means of pulling together the small dedicated computers into a cohesive network.

Project MIRACLE, which has been described in detail elsewhere by Beckwith (2), is also a hierarchical computer network. MIRACLE is an acronym for Multidisciplinary Integrated Research Activities in Complex Laboratory Environments. This real-time data acquisition system was started in 1971 under the auspices of a cost sharing grant (GJ-28455X) from the Division of Computing Research of the National Science Foundation and the Agricultural Experiment Station at Purdue. This system provides extensive use of modularly designed interfaces which in most instances avoid the costs of installing a small dedicated computer in the laboratory. These transmit digital data to a data concentrator computer. The latter acts as a storage buffer for the main processor. This multiprocessor system is in turn interfaced, via a high speed data link, to the extensive batch processing capabilities of the main campus computer. The laboratory user utilizes a teletype terminal that operates in a time-sharing mode from the main processor to initiate data collection from his laboratory instrument for automatic processing. This system has several advantages over a series of independent dedicated computers. The laboratory user need only concern himself with the development of the applications software, which may be written in a high level language. The modular interface is less expensive, easier to install and easier to operate than a stand-alone minicomputer. Further economies are achieved by the utilization of shared peripheral devices. The user has a large amount of computing power immediately available to him via his teletype terminal. The system is also designed to resist obsolescence and ensure the ease of future hierarchical expansion.

When the Natural Resources and Environmental Science Program started planning a laboratory automation and data acquisition system to serve several disciplines in the Agricultural Experiment Station at Purdue, it was confronted with the following problems:

1) The laboratory instruments were located in 9 different buildings and in two instances in remote research facilities 6–15 miles from the central campus.
2) The data generation rates and sampling durations of the analytical instruments spanned a wide range.
3) Some of the data acquisition problems required interactive capability with a computer while others did not.

4) Some of the potential users were relatively inexperienced in computer techniques.
5) Other users had already developed extensive software on the University's main batch computer to process data after it had been digitized manually and punched on cards.

Thus it was desirable to be able to utilize existing software without going through the effort of rewriting entire program packages for a different system. It was also desirable to be able to utilize the computing power of the CDC 6400-6500 for larger data processing jobs. Another feature we sought was the availability, for several of the research projects, of small reasonably priced computers that would also be compatible with a larger computer system; these would be used as preprocessors for instruments generating extremely large volumes of data to avoid overloading the central facility to the exclusion of other simultaneous jobs. Finally, budget limitations were a major factor that would not allow for the development of many individual dedicated computing systems.

With these constraints in mind the decision was made to design and develop a computer network with the expensive processor and peripheral equipment located centrally, smaller computers as nodes in the network, and with direct access to the CDC 6400-6500 via a high speed data link. After an extensive re-

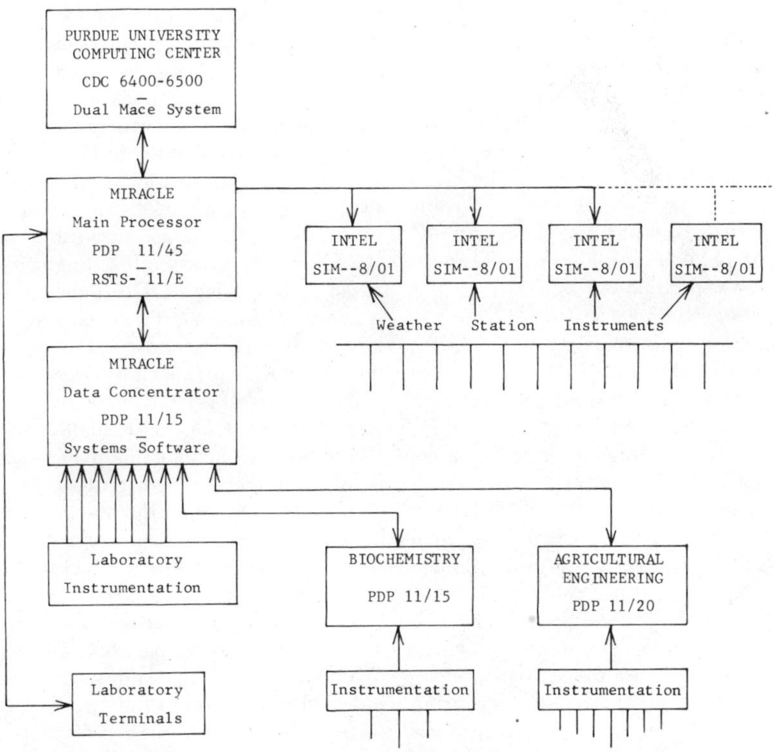

Figure 1. Current MIRACLE hierarchical system.

view, examination and evaluation of the numerous computing systems available, the PDP-11 system manufactured by Digital Equipment Corporation was selected.

The current configuration of MIRACLE is summarized in Fig. 1. It is a multiprocessor system utilizing four PDP-11 CPU's. The PDP-11/45 serves as the main processor utilizing the vendor developed RSTS-11 time sharing system with BASIC-PLUS language. Connected to this processor is a PDP-11/15 which acts as a data concentrator. Two laboratories have small satellite PDP-11 computers which serve as preprocessors for the high volume of high speed data generated in these laboratories.

In addition to the on-campus laboratories, there are 21 specialized micrometeorological weather stations located throughout Indiana. Selected stations will have Intel-SIM-8/01 microcomputers on site to acquire, integrate and store weather data for later transmission to the MIRACLE facility. Three of these systems have been fabricated, programmed and installed. Additional units will be installed at a later date. The MIRACLE system is also linked to the Purdue University Computing Center CDC 6400-6500 dual MACE system. The typical on-campus laboratory has a modularly designed interface which connects the laboratory instruments to the data concentrator. Data are transmitted from the instrument in serial digital form after passing through A/D and parallel to serial converters. In addition to the interface, the laboratory has a RSTS-11 teletype terminal.

The time sharing RSTS-11 system allows the researcher to start immediately to develop applications software in BASIC-PLUS with a minimum knowledge of computers or programming language. All laboratories may use the system simultaneously for a wide variety of tasks. From his laboratory terminal the user can enter, debug, run, store and retrieve programs. Utilizing his terminal he can initiate data collection from his laboratory instrument for automatic processing. If the data require extensive processing he may transfer the data file to the CDC 6400-6500 system. After he has logged on to the MIRACLE system, the user may then log on to PROCSY (Purdue Remote On-line Console System). His MIRACLE terminal then becomes a PROCSY terminal and the user has direct access to the computing power of the dual MACE system. The data link between MIRACLE and PROCSY is multiplexed, which allows seven simultaneous MIRACLE users to operate the PROCSY mode. After data are processed the researcher may request the output back to his own laboratory terminal or to any of the output devices of the two systems.

A considerable systems engineering and programming effort was involved in the in-house development of the laboratory hardware and systems software to allow the time sharing system to be used efficiently in a real-time data acquisition environment. The data concentrator (Fig. 1) is optimized to handle the fast response time for data acquisition that the time sharing system cannot provide. The optimization of the data concentrator software, written in assembly language, is based on task priority. The highest priority task is receiving the data from the laboratory instruments followed by data transfer to the time sharing system. Other tasks in order of priority are communication with the CDC 6400-6500 system, operating the incremental plotters, and finally a system status display.

The key to using a time sharing system in a real-time data acquisition environment is the buffering capacity of the data concentrator. As the data from the laboratory instruments enter the data concentrator they are temporarily stored in dynamically allocated buffers. Maximum buffer size is 32,768 16-bit words. Data in these buffers are accessed by user and system routines which transfer the data to the time sharing system for permanent storage and processing. Any over-loading of the buffers is prevented by the utilization of the two preprocessing PDP-11's in the laboratories that have either high volume or fast data rates. Data generation rates from the other laboratories do not impact the buffering capacity of the data concentrator.

The Intel-SIM-8008 microcomputers will be accessed directly from RSTS via voice grade telephone lines. The weather stations will be polled periodically and the data transmitted to campus for processing. These stations are buffered to store their data for as long as 29 hours before any overflow occurs.

An essential element in the successful implementation of the MIRACLE has been a dedicated professional staff operating the central facility. The group consists of a systems engineer who designed the interfacing hardware, an electronics technician who constructed these interfaces, and a systems programmer who wrote the necessary systems software needed to handle the flow of data from the laboratory instrumentation. The systems engineer also made major contributions in the development of systems software. Thus all systems engineering and systems programming has been completed by a professional staff; applications programming has been the responsibility of the research scientist.

The results of installing the MIRACLE system have not been a reduction in dollar costs or in personnel; in fact both have increased. What has happened is that the amount that a single research scientist and his students can accomplish in a given period of time has significantly increased. Data gathering and processing that formerly took months to complete are now done in hours or days. The neurophysiology group has the capability of scanning data in real time; this has resulted in fundamental changes in approach to describing nerve networks in mathematical models. A research group working with the gas chromatograph—mass spectrometer have discovered a new means of identifying complex compounds that would not have been possible without an on-line data processing system. The meteorological workers now have weather data analysis on a day-to-day basis rather than spending months on tedious manual processing of strip charts. The availability of timely weather data has resulted in bringing together a team of agronomists, plant physiologists, entomologists and agricultural engineers to develop a plant modeling and pest management strategy that can be updated daily. These are some of the examples of the innovations in research that have resulted from providing time sharing terminals in scientist's laboratories.

The authors wish to express their appreciation to D. C. Beckwith, W. H. Blake, F. E. Nay and V. N. Anderson for their efforts in the development of the systems software and hardware components.

REFERENCES

1. Ashenhurst, R. L. The minicomputer interfacing support system project. *Univ. of Chicago ICR Quart. Rept.* No. 33, 1972.
2. Beckwith, D. C. Computerization—A planned expansion. *Am. Lab.* 5 (2): 77, 1973.
3. Downer, S. W. A real-time, on-line analytical laboratory automation system with multiprogramming capability. *Am. Lab.* 6 (2): 93, 1974.
4. Epler, R. J. Minicomputers role in monitoring. *Environ. Sci. Technol.* 8 (1): 28, 1974.
5. Iracki, E., and H. V. Malmstadt. A computer-controlled automated system for chemical rate methods. *Am. Lab.* 6 (2): 56, 1974.
6. Klopfenstein, C. E. A planned data system. *Am. Lab.* 5 (2): 77, 1973.
7. Roberts, L. ARPANET: Current status, future plans. Networks for higher education. *Proc. EDUCOM Spring Conf.,* Washington, D. C., 1972.

Economies and economics of network computers

EDWARD K. BOWDON, SR.,[1] AND WILLIAM J. BARR

University of Illinois, Urbana, Illinois 61801 and

Bell Telephone Laboratories, Piscataway, New Jersey 08854

Network computers have evolved for reasons of economics and efficiency. A network computer is a set of computer systems which have been interconnected to achieve some common purpose: be it increased reliability, load sharing, or resource sharing. The major goal of most network computers presently in existence or planned is resource sharing (4, 10, 13). The operational and managerial advantages that may be derived from resource sharing and the attendant problems of network resource management are discussed in (14).

The following is a brief listing of some of the computer networks that have been or are being developed.

ALOHA—a satellite based network at the University of Hawaii.

ARPANET—a nationwide resource sharing network.

CANUNET—the Canadian University computer network.

CYBERNET—CDC's commercial network.

CYCLADES—the French connection.

DCS—an experimental local distributed computing system at the University of California at Irvine.

ICNS—the Integrated Computer Network System in England.

LIPS—the DEC Laboratory Interconnection Programming System.

MERIT—a regional educational network in Michigan.

NPL—a European computer network.

OCTOPUS—the Lawrence Livermore Laboratory general purpose scientific computing network.

TUCC—the Triangle University Computer Center network in North Carolina.

TSS—IBM's Time Sharing System network.

TYMNET—a national commercial network.

These network computers each offer one significant advantage that is not offered by single processor installations: the availability of large quantities of resources at a cost which is significantly less than would be required if each center purchased these resources individually. These resources are basically of two types: unique resources and common resources. *Unique resources* are those resources that, for whatever reason, can be made available only at a small number of locations (such as the array processing capabilities of the ILLIAC IV or the CDC STAR or specially developed software systems and proprietary programs). *Common resources* are those resources that are available at many centers. Increasing the common resources of a center offers the advantages of: *1*) improving the reliability of the computing services and *2*) facilitating load leveling (the process of transmitting tasks between centers to

---

[1] Present address: University of Arkansas, Fayetteville, Arkansas 72701.

U.C.W. ABERYSTWYTH LIBRARY

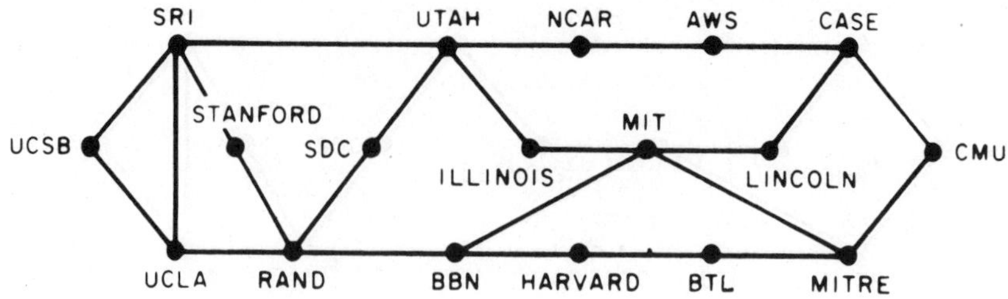

Figure 1. ARPANET.

improve the overall throughput of the network). From the user's point of view, load leveling improves the stability of the computing services that he uses since the peak demand periods at his home center can be smoothed into the slack periods at another center. These advantages, while enticing, remain advantages only if they can be achieved in a cost effective manner.

In the following sections we shall briefly examine four networks. These networks were selected for study because they are representative of the wide variety of philosophies and ideas that enter into the design of network computers.

ARPANET[2]

The Advanced Research Projects Agency Network (ARPANET), an ARPA research project, is the most widely discussed network computer and has probably provided the single most important impetus in the advance of network computer technology (3, 6, 11, 13). Furthermore, ARPANET is the largest fully automatic network computer in operation, both in terms of geographic distribution and number of centers. A representative topology for ARPANET, shown in Fig. 1, is a partially connected distributed network offering more than one path between any two centers.

The design of ARPANET was influenced by two major considerations: *1*) to explore network computer technology, and *2*) to interconnect those research centers across the country to which ARPA contributed substantial financing, thus allowing the sharing of unique resources.

One of the primary elements of cost in any network computer is topology. ARPANET is not a fully interconnected distributed network since this would imply direct paths between any two nodes. The most important consideration that led to the choice of a distributed network was reliability. Once it had been decided that ARPANET was to be a distributed network, considerable effort was spent in optimizing the final topology (9).

One of the most important side effects that resulted from the choice of a distributed network was the sophistication which is required in the Interface Message Processor (IMP). Owing to technological considerations, a distributed network that is not fully interconnected must use store and forward message switching. This requires considerable error-checking and decision-making capabilities in the IMP. For these reasons the ARPANET communications computer is significantly more expensive than that required for some other networks.

Available figures (private communication with Dr. M. Sher, Center for Advanced Computation at the University of Illinois) indicate that the basic IMP will run in the neighborhood of \$40,000 to \$50,000. However, the IMP by itself is

[2] See also paper by Frank Heart, p. 209.

not of much value, and from the beginning, work has continued on developing an intelligent terminal that can be connected to a Host. The initial work in this area resulted in a Terminal Information Processor (or TIP) which included the HP 316 IMP in its cost of about $92,000. As one TIP has ports for 32 terminals, several users with physical proximity may share the cost of a single TIP node in the ARPANET. However, the TIP has proved unsatisfactory for some applications and further work has resulted in the ARPA Network Terminal System (ANTS) which includes the HP 316 and a PDP 11. Work is now progressing to connect the $120,000 ANTS system to a CDC 6600 at Ft. Belvoir which will make the entire ARPANET look like simply an intelligent terminal (obviously one with large resources) to each Host. To become a node in ARPANET requires a minimum initial investment of at least $90,000. In addition, $20,000 to $35,000 in annual maintenance and ARPA privileges can be anticipated. At this time, only Federal agencies or Federal contractors can become ARPANET members; however, FCC approval has been given to two commercial organizations to set up national networks using ARPANET technology and to put user charges on a commercial basis.

What then are the motivations that justify the high cost of membership in ARPANET? Research oriented nodes are motivated to join ARPANET to gain the capability of using a wide variety of unique resources. Anyone doing work which requires large amounts of matrix arithmetic and array processing would clearly prefer to have the ILLIAC IV available than any of the currently popular general purpose processors. For many of the major universities, access to the ILLIAC IV alone may justify the expense of ARPANET membership. Another valuable unique resource is the University of Utah interactive graphics system. This likewise provides a significant amount of justification for membership. A network computer is a virtual necessity for these specialized and very expensive unique resources.

However useful the availability of these unique resources may be, for ARPANET to become a viable commodity without being subsidized, a wider variety of advantages must be offered. As ARPANET becomes more commercial, we shall see significant use of the common resources of the network. Since communication costs are expected to be low (about $0.30 per megabit after the first 4,000 megabits, which are included in the basic membership annual fee) load leveling on a network-wide basis may become feasible.

Perhaps the most radical possibility opened by ARPANET is that its successful implementation could lead to the elimination of most medium and large scale machines. The ANTS system clearly has the potential to operate as a remote job entry station for all kinds of tasks—including interactive procedures. Given this capability and assuming that ARPANET traffic capacity is great enough, we can envision a time when, instead of purchasing a medium or large scale computing system, the most economical action would be to purchase an ANTS and use already existing resources. This approach would probably require the use of brokers and would undoubtedly lead to the development of more specialized machines that could do certain types of tasks more cost effectively than general purpose machines.

THE TUCC NETWORK

The Triangle Universities Computation Center (TUCC) network is a system that is entirely different from ARPANET. The TUCC network is a regional (contained entirely within the state of North Carolina) homogeneous network consisting of an IBM 370/165 (at Research Triangle Park), an IBM 360/75 (at the University of North Carolina/Chapel Hill), and two IBM 360/40's (at Duke University/Durham and North Carolina State University at Raleigh). Also con-

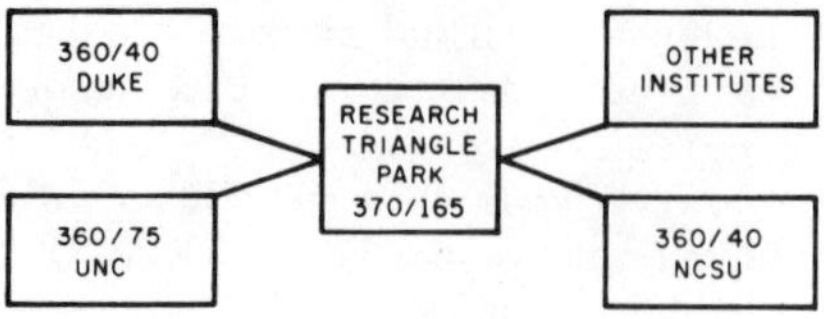

Figure 2. The TUCC network.

nected to the 370/165 are secondary terminals at about 40 other state educational institutions. These other institutions purchase machine time through the NCECS (North Carolina Educational Computing Service), which buys block time at wholesale rates from TUCC, and retails time to the other universities. In addition, NCECS provides consulting and other user services. The four processors are connected in a star topology, shown in Fig. 2, with a central message switch through which all center-to-center communications must pass. This topology was chosen because communication is expected to be mainly from the three universities to Research Triangle Park, and direct university-to-university communications are expected to be rare (12). Thus, extra links would be economically unjustifiable.

The total cost to the three TUCC universities for belonging to the TUCC network is in the range of $13,000 to $14,000 per month (15). While this seems high at first glance, this fee includes approximately 25% of the computing capacity of the 370/165. Adding this amount of capability to any of the centers individually would cost around $19,000 per month. Thus, if the capacity can be used, membership in the TUCC network is certainly economically justified.

The primary motivation in developing the TUCC network was economics. The 370/165 is primarily available to provide remote job entry and interactive services to the three universities. Most of the tasks processed at Research Triangle Park represent excess capacity demanded by users at the three universities. (This demand was sufficiently large enough to necessitate replacing the frequently saturated 360/75 with an IBM 370 in September, 1971.)

Two fairly unique advantages were obtained with the formation of the network. First, the universities became better able to attract faculty and staff who require large-scale computing services for research and teaching. A second important consideration was that TUCC was able to attract outstanding systems programmers, which the individual universities could neither afford nor keep busy.

Though currently operating in a cost effective manner, the TUCC network is far from becoming sedentary. Plans are currently under way to enable the 360's at the various universities to handle overflow from the 370. Obtaining this capability is especially important in a university environment where vacations, and hence due dates for term projects, can be staggered. While other services are provided by the TUCC network, the load sharing capabilities that are made available are sufficient to provide significant monetary savings, as well as improvements in service, to the three universities involved.

DISTRIBUTED COMPUTER SYSTEM

The Distributed Computer System (DCS) is planned as a local ring structure network, shown in Fig. 3, which is being designed and built at the University of California at Irvine (UCI). All communication is along the same path and there is only one path between any two centers. When completed, DCS is expected to provide medium and small scale computing services to the users at UCI. The following were the design objectives for DCS (8):

1) Reliability: The system should be invulnerable to total failure. The failure of any individual component should have minimal effect on the operation of the rest of the system ("soft fail").

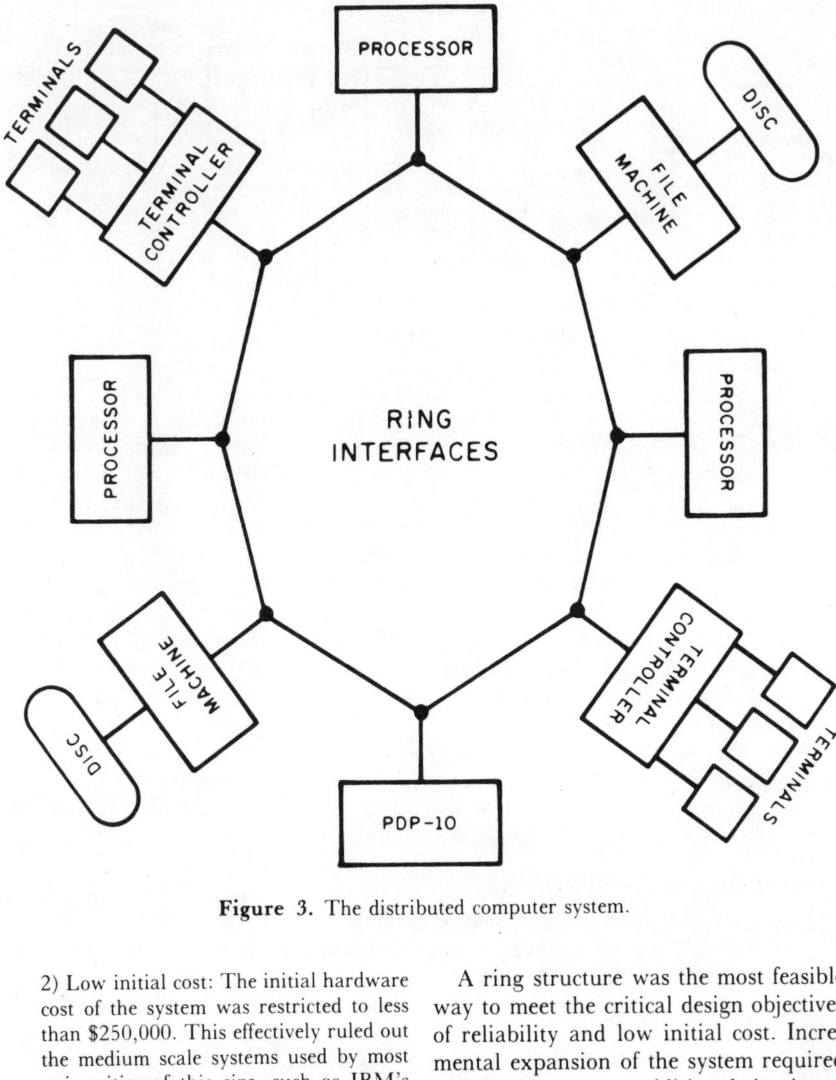

Figure 3. The distributed computer system.

2) Low initial cost: The initial hardware cost of the system was restricted to less than $250,000. This effectively ruled out the medium scale systems used by most universities of this size, such as IBM's 360/50, DEC's PDP-10, or XDS's Sigma 7. (At present a PDP-10 is being utilized in the initial stages of the work; its permanent inclusion is uncertain.)

3) Incremental expansion ability: The system should be capable of expansion of virtually any type, at a relatively low expansion cost.

4) Variety of language systems: Since this is to be a university computing system, a wide variety of computer languages and resources should be available.

A ring structure was the most feasible way to meet the critical design objectives of reliability and low initial cost. Incremental expansion of the system required only hooking up an additional ring interface and attaching it to the new processor. There is no theoretical limit to the expansion of the system, although in practice the number of processors the ring can efficiently handle is currently unknown. Expansion is economical because of the low cost of the ring interface (about $500) (7). A variety of language systems can be obtained by attaching small language-oriented processors to the ring. The system programming objective is achieved by

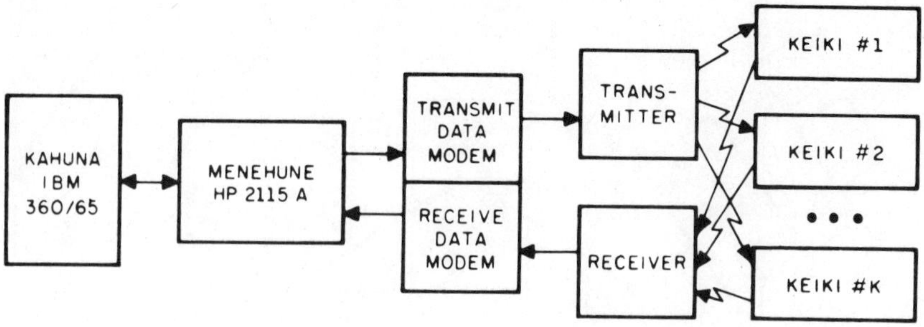

Figure 4. The ALOHA system.[7]

building a distributed operating system to maintain the soft-fail capability of the ring.

THE ALOHA SYSTEM

In each of the networks discussed so far, the communication media used has been wire. In the continental United States, where telephone networks already exist, or in local networks, wire is currently the most cost effective communications medium (6, 8, 13, 15). This is not the result of any inherent qualities in the wire, but is due primarily to convenience. One proposed system that has selected an alternate medium is the ALOHA System, shown in Fig. 4.

The ALOHA System is being built by the University of Hawaii and is intended to provide computing services to users at the University's main campus in the Manoa Valley on Oahu, a 4-year college at Hilo, Hawaii, 2-year community colleges on the islands of Oahu, Kauai, Maui, and Hawaii, and associated research institutes within a 200-mile radius of the University. Long-range plans include extending the communications system throughout the Pacific area, possibly even for ship-to-shore use. Some fast, inexpensive method is needed to connect an IBM 360/65 computer at the University's main campus, called the Kahuna, with remote access terminals (Keikis) at all of the above-mentioned locations (1, 2).

Among the alternatives to wire communication are radio and light waves. At present ALOHA is being built using radio communication, but research on the feasibility of light wave communication using light emitting diodes is also under way. Radio communication was chosen because it is inexpensive to use over large distances (in fact, radio messages could be beamed in by satellite from all over the world), the Keikis can be mobile, and a communications protocol can be established that takes advantage of the burst nature of messages from a single Keiki while ignoring all the inactive Keikis. This protocol is what makes ALOHA so interesting.

While the Kahuna is kept busy with computing tasks, an HP2115A named Menehune (after a legendary Hawaiian elf) serves as its link to the radio communication channel. It is Menehune's job to multiplex messages between Kahuna and a large, indeterminate number of users. Menehune uses two 100 kHz band width radio frequency channels, one for receiving messages from the Keikis, and the other for sending messages to the Keikis. This gives, in effect, a "party line" system.

Because of the anticipated usage pattern of the Keikis, the designers of the ALOHA System have chosen to allow communication in burst mode only. Under this constraint, Menehune can be said to have a capacity of about 24,000 bits per second. In practice, since message arrivals are virtually random, the capacity will be much less. For message packets of about 700 bits, the system can support between 140 and 160 active users if the transmission rate is about 2 packets per minute. (This is a significant number of users since they are all using the same channel (5).)

The investment needed to become part of the ALOHA System is anticipated to be one teletype and the $3,000 Keiki (a one-time cost) (1). This is a very reasonable amount since it allows the user access to the complete computing capabilities of the University of Hawaii. And, should the user keep the Keiki long enough, the average monthly cost will be low enough to allow almost anyone, regardless of location, to join the system.

CONCLUSIONS

Instead of looking at the technology of network computers, we have concerned ourselves with economic considerations. In the early days of the digital computer, there was widespread opinion that these machines were too expensive and too limited to be of much value to anyone but scientists and census takers. The situation has changed. Computer manufacturers are now subject to the same economic principles and constraints as most industries.

The key to cost effectiveness is efficient resource sharing since, without resource sharing, network computers have little of value to offer. Unfortunately, there is no formula for success in the construction of network computers; nor, however promising networks may be, can they be confidently expected to be the panacea for all problems. For certain high-data-rate laboratory work, a local mini may be the answer, either stand-alone or interfaced to a network. The available options are numerous and it is quite possible that networking will be the focus of innovative computer scientists and businessmen for some time to come. Given time, though, network computers will also enter the business world where good management and cost effectiveness reign as king and queen.

REFERENCES

1. Abramson, N. The ALOHA System. *ALOHA SYSTEM Tech. Rept. B72-1,* January 1972, 30 p.
2. Abramson, N. The ALOHA System—Another alternative for computer communications. *ALOHA SYSTEM Tech. Rept. B70-1,* ABRIO, 1970, 29 p.
3. Anon. *AFIPS Conf. Proc.* May 6-10, 1974, Montvale, N. J.: Am. Fed. Inform. Proc. Press, Vol. 43.
4. Anon. Digest of Papers from COMPCON 73. *IEEE Comput. Soc.*, Northridge, CA, 1973.
5. Bortels, W. H. Simulation of interference of packets in time ALOHA time-sharing system. *ALOHA SYSTEM Tech. Rept. B70-2,* March 1970, 26 p.
6. Cole, G. P. Computer network measurements: techniques and experiments. *Univ. of California, Los Angeles Sch. Eng. Appl. Sci. Report No. UCLA-7165,* October 1971, 350 p.
7. Farber, D. J. Progress report on the distributed computing system. *Univ. of California at Irvine Tech. Rept.*, January 1972.
8. Farber, D. J., and K. C. Larson. Supplement to proposal for research submitted to the National Science Foundation on distributed computing system. *Univ. of California, Irvine Tech. Rept.* October 1970.
9. Frank, H., et al. Topological considerations in the design of the ARPA net. *Proc. Spring Joint Comp. Conf.* 36: 581–587, 1970.
10. Greenberger, M. et al., editors. *Networks for Research and Education.* Cambridge, MA: MIT Press, 1974.
11. Heart, F. E., R. G. Kahn, S. M. Ornstein, W. R. Crowther and D. C. Walden. The interface message processor for the ARPA computer network. *Proc. Spring Joint Comp. Conf.* 36: 551–567, 1970.
12. Parker, L. T., Jr., T. M. Gallie, F. P. Brook, Jr., and J. K. Ferrell. Introducing computing to smaller colleges and universities—a progress report. *Comm. ACM* 12 (6): 319–323, 1969.
13. Roberts, L. G., and B. D. Wessler. Computer network development to achieve resource sharing. *Proc. Spring Joint Comp. Conf.* 36: 543–549, 1970.
14. Stefferud, E., D. L. Grobstein and R. P. Uhlig. Wholesale/retail specification in re-

source sharing networks. *Computers*, 6 (8): 31–37, 1973.

15. Williams, L. H. A functioning computer network for higher education in North Carolina. *Proc. Fall Joint Comp. Conf.* 41: 899–904, 1972.

Networks and the life sciences: the ARPA Network and Telenet

F. E. HEART
Bolt Beranek and Newman Inc.
Cambridge, Massachusetts 02138

The mushrooming of developments in computer/communication during the last 5 years has led to a rapid growth of interest in computer networks. A recent bibliography on Resource-Sharing Computer Networks (2) lists the number of published contributions for the years 1969, 1970, 1971, and 1972 as, respectively, 46, 94, 121, and 144. This sudden increase in interest was triggered by the United States' decision, in 1968, to build the Advanced Research Projects Agency (ARPA) Network. The ARPA Network has been widely reported, with many citations in (2); a few key references are (9–12, 15, 20) and recent interesting summaries of packet switching technology are given in (13, 19). Network projects are being undertaken all over the world; a recent document describing many such projects is (18), certain others are described in (4, 14, 17), and the involvement of satellite channels is discussed in (1). In the United States, commercial carriers based on the new technology are rapidly evolving (16, 22).

It has been possible for a long time for an individual to make use of several diverse remote computers; similarly, it has been possible to divide a problem among two or more geographically distant computers, or to use many different types of terminals with a computer service, and so on, and so forth. Such arrangements have typically been so laborious, however, that they have occurred very rarely. Now a new computer-communication partnership, and a burgeoning computer network technology, have moved a myriad of such activities from possible to convenient, and now such activities can and will really occur on a wide scale in many areas of application.

THE ARPA NETWORK

Figure 1 is a recent "logical" map of the ARPA network, showing Host computers as well as the network nodes (Interface Message Processors or IMPs, and Terminal IMPs or TIPs). By early 1974, internode traffic on this net had reached ~0.75 billion bits/day, and new nodes were joining the net at about one per month. Most internodal network circuits operate at 50 kilobits per second (kbs); however, two circuits operate at 230.4 kbs, the satellite link from the United States to Norway operates at 7.2 kbs, and the Norway-United Kingdom circuit operates at 9.6 kbs. Host interfaces to IMPs and TIPs typically operate at bit rates from 50 to 300 kbs, and most terminals are connected to TIPs at rates under 2,400 bits per second (bps).

Important features of the ARPA Network include the following.

It is a store-and-forward packet-switching system in which the path for a given message is not established in advance, each message carries an address, and messages typically traverse several nodes in going from source to destination.

It is a highly reliable and high speed communication system in which there are typically several possible paths between source and destination; packets are held at each node until successfully received at the succeeding node, and total source-destination transit time is usually under 0.5 sec.

The standardized IMPs/TIPs and the adaptive routing of packets make the net robust in the face of circuit or node failure and also permit extremely simple addition of circuits and nodes for expansion or reconfiguration.

The arrangements for Host connection facilitate the interconnection of computers from different manufacturers and of one-of-a-kind research computers.

The TIP provides access to the net, and in turn to the resource pool of Hosts, for a wide variety of terminal types. By use of a "network virtual terminal" mechanism, each network Host need deal with only that "virtual" terminal, all other terminal idiosyncrasies being dealt with by the TIP.

The basic way in which messages use the network capacity makes it appropriate to consider transmission costs to be independent of distance, and to consider such costs to be based primarily on quantity of data sent rather than connect time; both major departures from prior communication economics. Because circuit bandwidth may be quite flexibly and efficiently multiplexed, transmission costs are very favorable compared to other competitive approaches.

Programs running in the IMPs and TIPs may be remotely debugged, modified or reloaded from a Network Control Center, and a new version of the program can be released to the entire net in less than 2 hours.

The "Host-to-Host protocols" — disciplines by which processes in one Host

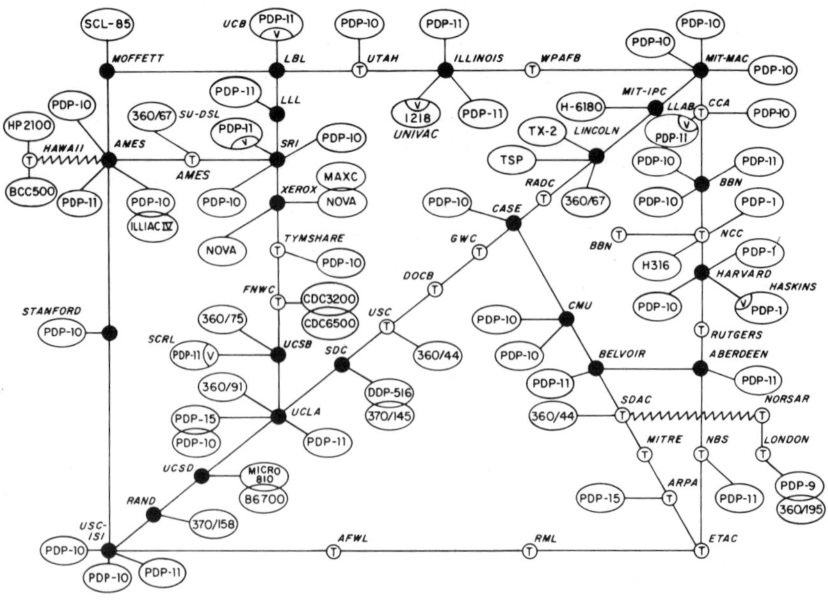

Figure 1. ARPA Network, logical map, May 1974.

communicate with processes in Hosts elsewhere on the network — are largely independent of the communication system, and a given Host grouping may implement and use a network standard protocol or may use a private specialized protocol. Thus the network conveniently allows diverse private "communities of interest" to be built up "on top of" the basic network functions.

As of May 1974, use of the ARPA Network is largely limited to ARPA-supported research groups and a few DOD agencies. There has been considerable interest in using the net by other branches of the government in general and by life science research groups in particular; to date the administrative hurdles have not been surmounted. It would be of considerable national benefit if the ARPA Network were made available to members of the life science community, especially toward the goal of their rapidly gaining experience with this new kind of resource sharing.

TELENET

Several organizations in the U.S. are attempting to offer packet communication service on a commercial basis as common carriers. The Federal Communications Commission has been receptive to this notion and several licenses have been granted for offering such service. The plans, status, and proposed price structures of one such organization, Telenet, provide a reasonable view of how this commercial service will appear, and are summarized below.

Telenet received FCC approval as a common carrier in April 1974, and plans to offer initial service in seven cities in early 1975. A map of the planned initial net is shown in Fig. 2; as shown, the circuits are planned as 50 kbs and 9.6 kbs terrestrial circuits, to make the initial net fully "3-connected." It is planned that the 9.6 kbs circuits will be upgraded to 50 kbs as traffic grows to more than 9.6 kbs per city. Telenet is next planning a rapid expansion to 18 cities and is also planning to employ domestic communication satellite services.

The Telenet IMP and TIP functions are being planned on equipment which permits easy adaptation of the ARPA Network software (which is in the public domain). However, a number of modifications of ARPA Network technology are planned, including: *1*) the siting of duplexed IMPs and TIPs in central offices (rather than on customer premises) and the use of manned central offices with backup power; *2*) a Host interface to IBM 360 and 370 equipment which is "native" and avoids changes to the Host software; *3*) implementation of a terminal interface to remote-job-entry equipment such as the IBM 2780.

The initial planned Telenet tariff is as follows, using packets of up to 1,000 bits:

PACKET CHARGE	$1.25/kilopacket
PORT CHARGE FOR LEASED ACCESS LINES:	
Up to 9.6 kbs	$50/month
50 kbs	$100/month
DIAL-IN PORT CHARGE:	
0–1,800 bps	$1.00/hour
2,400 bps	$2.00/hour
4,800 bps	$3.50/hour

It is intended that the high connectivity, manned offices, duplexed systems, and tested software will provide very high levels of reliability; and that the tariffs will provide economies to the user that will dominate other communications approaches. Notice the crucial facts that cost is independent of distance and based on the amount of data transmitted; quite a change in the commercial communication world!

OPPORTUNITIES

Economy

Networks are going to permit many straightforward but significant economies. For example, one U.S. university made the following approximate computation:

cost of existing campus computer center	\$1.5M/yr
cost of equivalent computer service from large machine 100 miles away	\$0.6M/yr
cost of network service to access remote machine	\$0.1M/yr
saving	\$0.8M/yr

This kind of thrift is a considerable incentive to determining the feasibility of employing the remote center; the cost squeeze on universities becomes worse every year and a good deal of pride can be swallowed for three quarters of a million dollars per year. (By way of partially explaining the surprising economy available by such routes, one should note that the total cost of a computer center, including building space, air conditioning, operators, and the like is usually about three times the cost of the computer in the center.) Of course, there are real problems — loss of control and potential lack of responsiveness to changing needs, priority conflicts, and so on — but many of these problems have been solved to a considerable degree in real cases. As an example, the University of Illinois has a sizable research and development group fully dependent on the ARPA Network for all its computing. As a second example, Harvard University has discontinued operation of a computer center in favor of obtaining service commercially from a local company, and MIT is also considering such approaches.

Competition

In many industries, it is competition that induces cost-effective performance. A competitive forcing function has been present in the minicomputer market, but has been usually absent in the domain of large computers, corporate/university computer centers, or service bureaus. All too often the user is effectively trapped by the following constraints: *1*) The university (or corporation) computer center is the only one to which the user is attached, all his files are there, and political obstacles may exist to his even considering

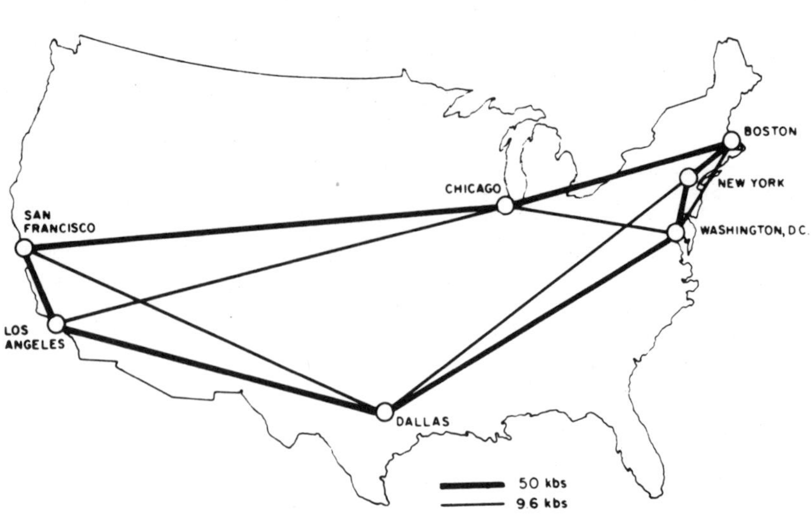

Figure 2. Telenet seven-city network.

outside services; and 2) The user is technically bound by terminal equipment and protocols that are unique to the entrapping computer.

Under such conditions, service may be very poor and very expensive, and there is little economic pressure for improvement. Once connected into a network, classes of users with straightforward computational problems are suddenly able to move files independent of distance, and to use the best, most cost-effective service on the net. There is immediate competitive pressure for improvement, easy comparison with alternatives, and a quantum increase in efforts toward good documentation, reliability, friendliness, improved rate structures, and the like. In the long run, such competition will be fed back to the manufacturers as well, and we can hope for survival of the fittest.

Expensive and specialized resources

There has been a tendency for science to require expensive facilities. In the computer field, such facilities include very large computers (e.g., ILLIAC IV), unusually large mass storage devices (10^{12} bits or more), specialized input/output devices (e.g., film readers, photocomposition devices), as well as large data bases, or large specialized bodies of software. At present the cost of such facilities is usually too great for any one group to afford, and the facility may quite literally not exist unless its cost can be distributed over the needs of many user groups. Network technology provides a convenient, highly cost-effective way to allow many geographically separated groups to share such computer resources, and the network is therefore a major factor in whether the facility can be afforded at all.

In the life sciences the possibilities abound, including: large data bases (collections of literature, drug data, public health data), specialized software systems to support research in particular fields (5–8) (for instance EKG/ECG analysis systems, biological modeling facilities, chemical analysis/synthesis systems), and powerful bibliographic systems.

More generally, most computer systems embody hardware and software that is inherently difficult to transfer to other settings; obtaining use of someone else's program and system by using it remotely on its home ground is a very important new approach.

Modified roles for small machines

In many laboratory settings, a small computer with analog-to-digital conversion equipment, modest mass storage, and displays is virtually standard equipment. The advent of networks has two distinct potential impacts on this arrangement. First, it may be feasible in some instances to reduce such instruments to still smaller "terminal-like" devices and then to access larger machines on the net for mass storage, common software packages, and other assistance. In some cases such a step would be most welcome in reducing the burden of running a computer center. Biomedical laboratories in particular may view a medium size computer installation as quite a diversion, involving funding and engineering staffing that may be hard to locate and retain.

Second, networks will permit an important and novel way to expand the utility of small local computers, by sending programs from a central distribution point for actual running in the small local machine. The ARPA Network has shown that with reliable communication, it is perfectly feasible to load new code into distant machines, and this approach may permit tailoring a small local machine to the needs of the moment. In both approaches, more power is obtainable with a smaller local system.

Human–human interaction

Once the ARPA Network had become operational on a widespread basis, a relatively simpleminded idea that, in rudimentary form, had been a curiosity at some installations for years, suddenly and quite unexpectedly became a very successful fringe benefit. The current form of the system on the ARPA Network is called "network mail," and provides that many individuals have computer "mailboxes" in their favorite Host computer; whenever A desires to communicate a message to B, C, D, and E, he logs into a convenient Host, uses a program that sends mail, lists the addressees with an indication of the addressees' mailboxes, and types the message. No matter where B, C, D, and E might be, they occasionally enter the net at a local node, log into their (mailbox) Host, and as part of the log-in process are automatically told that a message exists for them. They may then obtain that message or any other pending messages, using a program to read their mail.

For significant numbers of people, this system has become, almost overnight, an essential aid to their work. A directory (3) issued in January 1974 included more than 450 individuals with network addresses, and the actual number having addresses and using the facility was probably several thousand. The system has many elaborations, but its primary impact has been to permit communication at any hour of the day or night without even knowing the city, state or country where the addressee might currently be traveling.

Other similar systems for human–human interaction are being designed, including various forms of teleconferencing (21), and it is easy to imagine these new approaches to human communication becoming very popular among the academic and industrial communities.

Document preparation

Another kind of human–human interaction can be seen emerging from the already widespread use of systems for on-line text editing and document preparation. The notion of collaborative work on a paper by people who are geographically distant becomes obvious, and the "turnaround" time becomes minutes rather than days for new drafts. Further, there is a strong incentive for the development of high quality final output printing systems, spelling-correction systems, grammar-assistance systems, and so on. A number of documents have been produced by "distant" groups of people, using the ARPA Network, and the interaction has apparently been very convenient and stimulating. Within the ARPA community a major effort is now being mounted to pursue these kinds of tools for expediting human interactions.

REFERENCES

1. Abramson, N. Packet switching with satellites. *AFIPS Conf. Proc.* 42: 695–702, 1973.
2. *Annotated Bibliography of the Literature on Resource Sharing Computer Networks.* Washington, D.C.: U.S. Department of Commerce, National Bureau of Standards, NBS Special Publication 384, SD Catalog No. C13.10:384, Sept. 1973.
3. *ARPA Directory.* NIC 19275, Network Information Center, Menlo Park, Calif. Jan. 1974.
4. Barber, D. L. A. The European computer network project. *Proc. Intern. Conf. Comput. Commun.* ICCC 72 CHO 690–8C, Oct. 1972, p. 192–200.
5. Buchanan, B. G., and J. Lederberg. The heuristic dendral program for explaining empirical data. *Proc. IFIP Congr. 1971* (Ljubljana), Aug. 1971, p. 91–100.
6. Castleman, P. A., C. H. Russell, F. N. Webb, C. A. Hollister, J. R. Siegel, S. R. Zdonik and D. M. Fram. The implementation of the PROPHET system. *AFIPS Conf. Proc.* 43: 457–468, 1974.
7. Corey, E. J., and W. T. Wipke. Computer-assisted design of complex organic syntheses. *Science* 166: 178–192, 1969.
8. Feigenbaum, E. A., B. G. Buchanan and J. Lederberg. On generality and problem solving: A case study using the dendral program. In: *Machine Intelligence 6.* Edinburgh, Scotland: Edinburgh Univ. Press, 1971.

9. HEART, F. E., R. E. KAHN, S. M. ORNSTEIN, W. R. CROWTHER AND D. C. WALDEN. The interface message processor for the ARPA computer network. *AFIPS Conf. Proc.* 36: 551–567, 1970.
10. HEART, F. E., S. M. ORNSTEIN, W. R. CROWTHER AND W. B. BARKER. A new minicomputer/multiprocessor for the ARPA network. *AFIPS Conf. Proc.* 42: 529–537, 1973.
11. MCKENZIE, A. A., B. P. COSELL, J. M. MCQUILLAN AND M. J. THROPE. The network control center for the ARPA Network. *Proc. Intern. Conf. Comput. Commun,* ICCC 72 CHO 690-8C, Oct. 1972, p. 185–191.
12. MCQUILLAN, J. M., W. R. CROWTHER, B. P. COSELL, D. C. WALDEN AND F. E. HEART. Improvements in the design and performance of the ARPA network. *AFIPS Conf. Proc.* 41: 741–754, 1972.
13. METCALFE, R. *Packet Communication*. Project MAC Report #MAC TR-114, MIT, Cambridge, Mass., Dec. 1973.
14. OIZUMI, J. Plans for a Japanese university computer network. *Proc. 7 Hawaii Intern. Conf. on System Sci.* (*Comput. Nets Suppl.*), Western Periodicals Co., Jan. 1974, p. 27–29.
15. ORNSTEIN, S. M., F. E. HEART, W. R. CROWTHER, H. K. RISING, S. B. RUSSELL AND A. MICHEL. The terminal IMP for the ARPA computer network. *AFIPS Conf. Proc.* 40: 243–254, 1972.
16. PACKET COMMUNICATIONS INC. *Application to the FCC for Common Carrier Status.* FCC No. P-C-8533, Jan. 1973.
17. POUZIN, L. Presentation and major design aspects of the Cyclades computer network. *Proc. Datacomm 73,* IEEE 73 CHO 828-4C, Nov. 1973, p. 80–85.
18. *Proceedings of the International Advanced Study Institute on Computer Communication Networks.* Leyden, The Netherlands: Noordhoff, 1974.
19. ROBERTS, L. G. Data by the packet. *IEEE Spectrum* 1974, p. 46–51.
20. ROBERTS, L. G., AND B. D. WESSLER. Computer network development to achieve resource sharing. *Am. Fed. Inform. Proc. Soc.* (*AFIPS*) *Conf. Proc.* 36: 543–549, 1970.
21. Several Papers, Session on Teleconferencing. *Proc. Intern. Conf. Comput. Commun.* ICCC 72 CHO 690-8C, Oct. 1972.
22. TELENET COMMUNICATIONS CORPORATION. *Application for a Public Packet Switched Data Communications Network.* FCC No. P-C-8750, Oct. 1973.

CRYSNET

a crystallographic computing network with interactive graphics display[1]

E. F. MEYER, JR., C. N. MORIMOTO, JOSÉ VILLARREAL, H. M. BERMAN, H. L. CARRELL, R. K. STODOLA, T. F. KOETZLE, L. C. ANDREWS, F. C. BERNSTEIN AND H. J. BERNSTEIN

Texas A&M University, College Station, Texas 77843
Institute for Cancer Research, Philadelphia, Pennsylvania 19111
and Brookhaven National Laboratory, Upton, New York. 11973

More than the simple study of crystals, modern crystallography can be defined as the study of the structure of crystalline matter at or near atomic resolution by means of diffraction of X-rays, neutrons or electrons (15). As a science, crystallography throbs with relevancy (8). Crystallography settled the debate over the chemical form of NaCl (1914), served as a basis for the understanding of the chemical bond (1938), and more recently ushered in the now widely popular field of molecular biology (1953). Within the past 10 years the number of small compounds under study has risen to a current level of about 2,000 reports per year, and equally important, the number, size, and variety of macromolecules under investigation is steadily increasing.

The computing needs of crystallography can be classified in terms of data preparation, structure solution (phasing and electron density maps), structure refinement, and information storage and retrieval. It is sufficient for our present purposes to establish that computational needs are essential (5) and substantial.

The computational tasks in crystallography are worth considering here because, as a complement to the seminal ideas, advances in computing (and instrumentation) have been essential to the study of a host of materials. To gain an estimate of the number of data points measured and treated computationally, consider that small molecule studies may use from several hundred up to 5,000 independent intensity values. A low resolution data set of a protein may involve less than a thousand measurements. As macromolecular studies approach the level of resolution permitted by the crystalline material, the number of measurements for one isomorph can exceed 10,000 intensity values. Each of these values contributes to the calculation of

[1] Work performed under the auspices of the U.S. Atomic Energy Commission and supported by the National Science Foundation (AG-370 and GJ-33248X) and in part by the National Institutes of Health (CA10925 and RR05539). This manuscript was distributed and edited via CRYSNET.

electron densities at each of, for example, 10^6 grid points ($100 \times 100 \times 100$ grid) in an electron density map. The simple number of additions and multiplications required is then some multiple of 10^{10}. Yet new computational advances allow the production of an electron density map in the order of minutes on a large computer.

Recent hardware advances have permitted numerical information to be presented in graphical form (12). Human perceptions are usually better at processing pictorial information than lists of numbers or characters; "A picture is worth a thousand words." How can we describe the complexitites and interactions of the active site of an enzyme in a quantitative sense? A graphical language is required. If the user needs to be able to change (interact with) the picture then we have established a need for interactive computer graphics.

The scientific community as a whole currently requires techniques to aid the interpretation of final crystallographic results and their application to other areas of science. Unfortunately, with the exception of the study of signal molecules (e.g., penicillin, vitamin B_{12}, and so on) or macromolecules, the fruits of the crystallography of organic and biological materials are most manifest to the scientific community in terms of model kits. How else can a chemist construct ball/stick models with considerable assurance of representing the stereochemistry and architecture of compounds of interest? Yet, the synthesis of chemical information from increasingly larger data bases and the facile modeling of increasingly elaborate molecules to the end of understanding biological processes at the molecular level challenge man to extend his perceptions.

For the practicing X-ray crystallographer, computational requirements are compounded by the need to see results at various stages. He must prepare and interpret three-dimensional representations of electron-density maps. These maps most commonly look like stacks of weather maps with isobars representing levels of electron density. Atoms are found at peaks in the maps. Finally, a fully resolved picture is obtained from the measured X-ray diffraction intensities and their elusive phases. Then as the model takes form, small shifts in the positions of the atoms, together with estimates of their thermal motion, can be adjusted to give an optimum fit between calculated and measured intensities. Through this refinement process for small molecules, distinctions in bond type and order can be derived from the molecular geometry (e.g., a C=C double bond at 1.337 ± 0.006 Å is shorter than a C—C single bond, which is typically 1.541 ± 0.003 Å).

The final results from individual investigations have been collected in two principle data bases: the Crystal Data File (1) for small molecules, and the Protein Data Bank (ms in preparation) for macromolecules. The use of interactive graphics permits the facile survey of these data bases by means of information retrieval systems such as that implemented by Feldmann (7) at the National Institutes of Health or the programs under development in CRYSNET (3) like Search (11).

In the biological sciences, it is the exception to find a laboratory with one or more Kendrew (scalewise) models of macromolecules. The new Protein Atlas (13) has anticipated a fair number of questions in its presentation of the architecture of macromolecules. Unfortunately, both the primary structural data (atomic coordinates) and secondary data (Kendrew models) are not palatable to many in the biological sciences. The Protein Atlas or approximate models like those of Labquip[2] serve a qualitative function in presenting the topology of the macromolecule. For a quantitative presenta-

---

[2] Labquip, 18 Rosehill Park Estates, Caversham, Reading RG4 8XE, England.

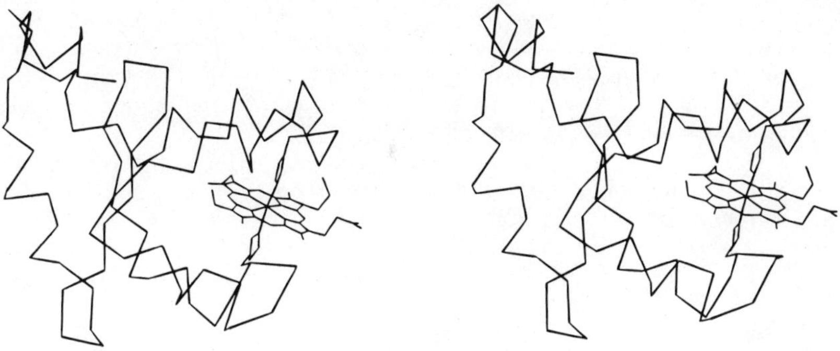

Figure 1. Stereo view of cytochrome b_5 (10) showing the haem group, His-39, His-63, and the α carbon backbone. Prepared on the CRYSNET Vector General display from an entry in the Protein Data Bank.

tion of the molecule, interactive computer graphics can serve as the "eyepiece" to a selective "microscope" that can survey large data bases, reduce the information to a pictorial level, and present the view from a variety of vantage points. Figures 1 and 2 illustrate the stereo effect of the motion picture film shown at this conference. In addition to presenting a picture the computer can answer questions about the geometry of the molecule under study within the limits of resolution of the primary data. The analogy with the microscope can be carried one step further because of the emerging prospects for computational modification of molecules to mimic chemical reactions. The investigator could establish a graphical experiment and follow or alter its course visually by means of the display. An example of this might be the introduction, docking, and quantitative evaluation of a substrate or inhibitor molecule within an active site of an enzyme molecule.

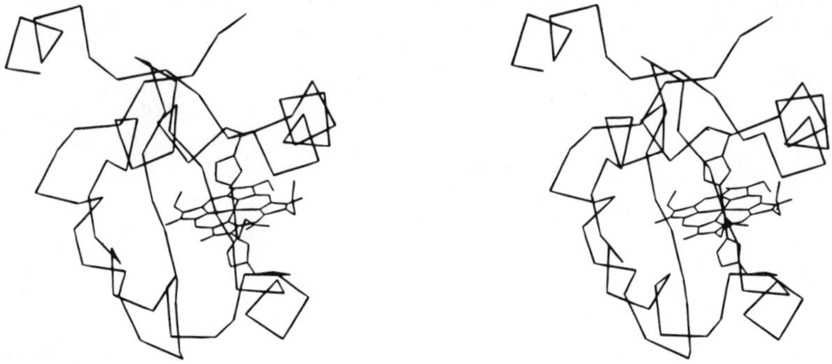

Figure 2. View of Fig. 1 rotated 45°. Viewing these stereo images may be facilitated by placing a card between them. A plastic viewer may be purchased from Hubbard Scientific Co. P. O. Box 105, Northbrook, Illinois 60062. These views, although constrained by the printed page, attempt to illustrate the kinetic depth effect of real-time rotation of the displayed image.

Traditionally, scientific computing has been performed on a local computer in the "batch" mode, in which a deck is submitted and output eagerly awaited. Batch processing is a viable and computationally efficient process, except for the considerable duplication of effort in program development and maintenance. However, two new trends are now emerging to challenge the local centers. The one is the laboratory minicomputer, which might operate one or more instruments, perform initial data correction operations, and control a graphics output device. The other trend stems partly from the nature of the problems at hand. Certain jobs require conversational computing that may not be supported locally. More telling still is the case in which the problem greatly exceeds the capacity of the local computer. By persistence one may manage to travel to Boulder or Berkeley or Brookhaven to use a CDC 7600, or else one must learn to think "small." These two trends merge in the intelligent terminal and the computer network. The intelligent terminal has sufficient computational power to permit interactive processing on a "stand alone" basis. In addition, it may be linked to one or more giants of the computer industry that may house (and process) large data bases or permit multiparameter processing.

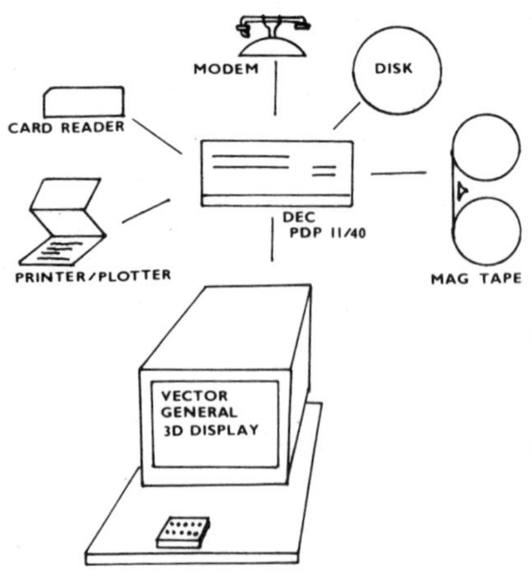

Figure 3. Schematic drawing of the CRYSNET terminal configuration.

The CRYSNET system is a network of three original nodes, *1*) Brookhaven National Laboratory, New York; *2*) Institute for Cancer Research, Philadelphia; and *3*) Texas A&M University, together with a new node without graphics capability at *4*) Johns Hopkins University. These terminals are linked to the Central Scientific Computing Facility at Brookhaven National Laboratory by means of direct distance dial (voice-grade) telephone lines. There are currently two CDC 6600 computers at Brookhaven with a 7600 computer scheduled to be installed later this year.

The intelligent terminals (1–3) each consist of (Fig. 3):

- PDP 11/40 computer with 28k words of memory
- 1.2 M word disc pack
- magnetic tape transport
- card reader
- line printer
- Vector General graphic display with 3-D rotation hardware
- communications interface, and modem.

Communications are transacted at 2,000 baud over synchronous, halfduplex lines using a CDC mode 4C protocol.

Work on CRYSNET is entering its third year. Many of the users are also programmers. Most of the programming effort has been directed toward *1*) consolidating and documenting the standard crystallographic programs at Brookhaven, adapting these programs to allow convenient remote use, and *2*) interactive display of molecular structures. Under development are electron density map contouring and model-fitting programs as well as information retrieval programs.

The advantages and disadvantages of networks and resource sharing have been discussed (2, 6, 16). It is curious that of the few major crystallographic groups with access to the ARPA net, none have used it. Our experience underscores several easily recognized advantages to CRYSNET. Central site maintenance of a program library and program documentation, together with the facility to trace major bugs, is a happy advantage. More significant perhaps is the mutual cooperation of several groups with individual interests. Hardware compatibility and occasional personal contact contribute to the program at each site during the developmental stage. Telephone costs and occasional disruptions (over low grade phone lines) are one hindrance for long distance operation and the problem of hardware maintenance is another.

The initial cost of each PDP 11/40—Vector General system is about \$70,000. A roughly equal amount has been spent on the salaries of the individuals involved in the original effort. As this effort approaches fruition the costs are being balanced by the utility of the crystallographic program library and by the superior capability of the computer graphics for model building and study. Novel methods (9, 14 and D. M. Collins, ms submitted for publication) for solving or refining macromolecular structures are dependent on large computational capacities like those provided by CRYSNET. Several of the new methods promise to require considerably less effort in the laboratory for preparation of heavy atom derivatives and the like with accompanying savings in resources.

Costs for telephone time vary from day rates of about \$0.45/minute to night rates of about \$0.25/minute. Transmission demands may be considerably greater at the program development stage than at the production stage. From the F.C.C. to the courts to the free (?) market economy, the entire subject of telephone data transmission is in a state of flux. Several links (AT&T, Datran, etc.) now exist between major cities. These firms project future data transmission demands that appear astronomical.

The linear decrease in cost of minicomputers over the last 5–6 years (but not peripherals!) and the incremental increase in capacity with lowered user costs of present (IBM 370/195, CDC 7600, etc.) and future (TI ASC, CDC STAR, etc.) generation large computers present two clear trends in the computational market place. A division of labor can be established in a network environment. Yet, networks like CRYSNET may be the victims of their own success. Why so?

First, funding agencies appear eager to support promising projects on a trial basis but loath to support permanent research facilities. Second, the benefits of a successful network might be less accessible to the small user than to the large. The large computer user now has found his own modus vivendi, so that only when the necessities of economics or capacity dictate a change from traditional in-house batch processing is he likely to add support to the networking concept. Yet, without his support the successful implementation of a dedicated network such as CRYSNET cannot make a real impact on the scientific community. An exception to this is being felt in subtle ways: the joining of minds in a closely knit networking environment has demonstrated in a few signal cases that the whole is greater than the sum of the parts. A critical mass is reached. Unexpected results occur. A new avenue for creative processes is opened. A synthesis of minds and interests creates new possibilities for creative thought in science. And interactive graphics lets us use the computer as a research tool, but on our terms.

Specifically, the Austrian physicist, Ernst Mach (4) formulated the principle of economy in science: the scientist must seek the simplest means of arriving at his results and must exclude elements not perceived, directly or indirectly, by the senses. An object of these efforts is to let the scientist extend his perceptions to the molecular level.

Until recently, the capacity for interactive computer graphics to display molecular models has been limited to less than 10 laboratories in the world, with hardware costing several hundred thousand dollars apiece. Now, computers and graphic displays are "off the shelf" items, still expensive, but considerably more accessible, and the trend is favorable. Alone, this is not sufficient. However, when the "microscope" function of interactive computer graphics is evoked, the potential impact on the biological sciences should be clear. How do you create models of viruses or membranes when high resolution structural information becomes available? How do you describe the geometric changes that produce information transfer in enzyme complexes? How do you perceive the qualitative interaction of small molecules with receptor sites? An intelligent, stereographics terminal with access to computing power and data bases may well hold the answer.

REFERENCES

1. Allen, F. H., O. Kennard, W. D. S. Motherwell, W. G. Town and D. G. Watson. Cambridge crystallographic data centre. II. Structural data file. *J. Chem. Doc.* 13: 119–123, 1973.
2. Aufenkamp, D. D., and E. C. Weiss. NSF activities related to a national sciences computer network. In: *Networks for Research and Education: Sharing Computers and Information Resources Nationwide*, edited by M. Greenberger, J. Aronofsky, J. L. McKenny and W. F. Massy. Cambridge, Mass.: MIT Press, 1973.
3. Bernstein, H. J., L. C. Andrews, H. M. Berman, F. C. Bernstein, G. H. Campbell, H. L. Carrell, H. B. Chiang, W. C. Hamilton, D. D. Jones, D. Klunk, T. F. Koetzle, E. F. Meyer, C. N. Morimoto, S. S. Sevian, R. K. Stodola, M. M. Strongson and T. V. Willoughby. Second Annual AEC Scientific Computer Information Exchange Meetings, Proceedings of the Technical Program, 1974, 148–158.
4. Bradley, J. *Mach's Philosophy of Science.* London: Athlone, 1971.
5. *Computational Needs and Resources in Crystallography*. Proceedings of a Symposium, Albuquerque, New Mexico, April 8, 1972, sponsored by National Academy of Sciences, Washington, D. C.
6. *Computational Needs and Resources in Crystallography* (ref (5)), p. 105–108.
7. Feldmann, R. J. Interactive graphic chemical structure searching. In: *Computer Representation and Manipulation of Chemical Information*, edited by W. T. Wipke, S. R. Heller, R. J. Feldmann and E. Hyde. New York: Wiley, *Interscience* 1974, 55–82.
8. Hamilton, W. C. The revolution in crystallography. *Science* 169: 133–141, 1970.
9. Klunk, D. G., and E. F. Meyer. Application of Boolean programming optimization methodology to the direct solution of crystal structures. *Nature* 248: 761–762, 1974.
10. Mathews, F. S., M. Levine and P. Argos. Three-dimensional Fourier synthesis of calf liver cytochrome b_5 at 2.8Å resolution. *J. Mol. Biol.* 64: 449–464, 1972.
11. Meyer, E. F., Jr. Storage and retrieval of macromolecular structural data. *Biopolymers* 13: 419–422, 1974.
12. Newman, W. M., and R. F. Sproull. *Principles of Interactive Computer Graphics*. New York: McGraw-Hill, 1973.
13. Richards, F. M., and H. W. Wyckoff. *Atlas of Molecular Structures in Biology, Vol. 1, Ribonuclease S*. Oxford: Oxford Univ. Press, 1974.
14. Sayre, D. Least-squares phase refinement. *Acta Cryst.* A30: 180, 1974.
15. Stout, G. H., and L. H. Jensen. *X-ray Structure Determination—A Practical Guide* London: Macmillan, 1968.
16. *A Study of a National Center for Computation in Chemistry,* Division of Chemistry and Chemical Technology, National Aademy of Science—National Research Council, Washington, D. C. 20418.

Hierarchical computer systems for on-line applications[1]

R. L. ASHENHURST
University of Chicago
Chicago, Illinois 60637

Experiments in the physical and biomedical sciences are becoming increasingly dependent on digital systems for on-line control and data gathering. This mode of operation has received enormous impetus from the advent of the minicomputer, functioning as a flexible and inexpensive control device. Although much has been accomplished in the development of efficient facilities for the off-line processing and analysis of experimental data, a comparable effort has been slow to develop for on-line digital systems forming part of an experimental setup.

The need for such support facilities stems from the fact that an experimenter normally has need for a higher level of computer system than that required to run the experiment in routine operation, but this need is only occasional. During an experiment in progress it may be desired to transfer accumulated data in bulk to a more permanent storage repository, or make program modifications. Before an experiment begins there is a substantial amount of effort invested in testing experimental equipment and preparing control programs. These functions can all be expedited by the presence of expanded computer facilities. There is in addition the problem of maintaining adequate backup for malfunctioning computer equipment. What is at stake here is often much more than just the inconvenience of having costly experimental equipment idle while the computer is down.

In a wider area, applications of computer technology to clinical care have been much proposed, and in less measure successfully implemented. The most successful have been those that embody aspects of real time control, analogous to those just discussed for experiments. The least successful, on the other hand, have been those that attempt to superimpose computer systems on human functions, in the physician–nurse–patient system. It is a truism that such systems, in order to succeed, must gain acceptance by the physician, the nurse, and the patient, but in fact a formula for gaining such acceptance is still elusive. Although concepts of computer-aided diagnosis and computer-directed scheduling conjure up an image more or less exciting to various segments of the health care professions, a realistic view is that the more mundane objective of having appropriate information accessible to physicians and nurses so that they can use it in the practice of their professional skills in the way they have been trained is still far from achieved. Reasons

[1] This work supported by the National Scence Foundation, under grant GJ-33084X.

for this cover a spectrum of identifiable causes: systems are proposed that are too comprehensive and complex to be implementable at all; systems are proposed that are implementable, but the right mixture of medical and computer expertise is not applied so they fail: systems are proposed that are implementable and are implemented, but they are not institutionalized and so pass into limbo when certain key individuals disassociate themselves from them.

Since the effectiveness of much biomedical research is critically dependent on the taking of data in a clinical context, it follows that the development of clinical care information systems is also relevant to the application of computers to research methodology.

The objective here is to provide a computer system environment where basic information systems for clinical care can be built up, in such a way that increasing sophistication in automatic monitoring and evaluation can be incorporated as evolved. It is required that such systems be composed of modules which are under the control of those who run the operation that they support, that they both be able to function with existing systems which are not computerized, and be able to be interfaced with other computerized portions as they come into being.

Modern computer systems are technologically capable of monitoring the collection of data from dispersed points, assembly and storage at one or more centralized locations, transformation by computational processing of one sort or another, and distribution of results selectively to dispersed points. The assembled data can also be subjected to further processing for statistical summaries and clinical research studies. But this potential can only be realized if an appropriate information system design is superimposed on an appropriate computer system configuration. To do this it seems necessary, even if not always sufficient, to have the collaboration of computer systems professionals and biomedical professionals.

COMPUTER SYSTEMS TECHNOLOGY

Two technological developments in the computer field are relevant to providing an environment where computer support is available for on-line processes, both of the real-time instrumentation variety and where interaction with humans is required. The first is the introduction of the minicomputer, which allows systems to be run and controlled locally to their application. The second is the development of networks of computer systems, which permit sharing of program packages and data bases. The environment pictured here is one in which *1*) basic systems providing operational support are implemented in local minicomputers, running independently; *2*) these minicomputers, however, are interfaced to a hierarchical system providing backup services, which includes one or more large centralized general-purpose facilities in which program packages and data bases reside; *3*) these backup facilities allow for continuing development of the programs that run in the minicomputers, without the problems that usually accompany program development on isolated minicomputers; *4*) there is associated with the hierarchical system and network a staff of individuals with professional competence in both computer systems and biomedical applications, who are dedicated to the task of facilitating the operation of research and clinical services in the computer systems environment.

It is convenient to distinguish four major subsystems of a computer system operating as described: *a*) file subsystem; *b*) communication subsystem; *c*) processing subsystem; and *d*) human–machine interaction subsystem. The first three of these are separate functions within the computer system (by "communication" is meant data transmission, not intelligibility); the fourth is the interface of the computer system with the environment in which it must operate.

The evolution of the technology in the last few years seems to support two contrasting system trends: the possibility of connecting terminals to centralized computer systems, and networking these systems together, has made it possible to achieve more versatility in the "centralization" of computer capabilities, but the introduction of the minicomputer and inexpensive digital interfacing circuitry has made it possible to achieve more versatility in the "localization" of such capabilities. Each of the subsystem functions cited above has aspects more effectively handled by central general-purpose facilities, and aspects more effectively handled by local dedicated facilities.

These aspects can be summed up by applying the words "reliability" and "flexibility" in two different ways. Reliability can mean dependability for systematic operation according to fixed standards (centralized systems), or it can mean availability at times most convenient to users, and under their control (localized systems). Flexibility can mean capability of being adapted to a wide variety of tasks performed according to general system standards (centralized systems), or capability of being modified over short time periods in response to user-specified requirements (localized systems).

THE HIERARCHICAL APPROACH

The Minicomputer Interfacing Support System (MISS) developed at the University of Chicago effectively allows advantages of centralized and localized systems to be combined, with a minimization of the disadvantages. Small dedicated online systems can be developed on minicomputers, which when operational are under the control of the departments for which they function, but these systems can interact with general backup systems, both in their normal operation (mass data transfer, off-line processing) and during development (programming assistance, testing).

An important concept of MISS is that of the intermediate hierarchical level. All that can be said about combined local-central computer systems is at least in principle implementable by connecting local dedicated systems directly to a central general-purpose facility. But here the interests of the local system user and the central system management tend to come directly into conflict, as many have found who have tried to use such a configuration. By interposing an intermediate level, under control of a group dedicated not to the individual applications but to the support of such applications (as contrasted, e.g., to large-scale numerical computing or interactive artificial intelligence), this conflict is mediated. The issues involved in a three-level versus a two-level hierarchical systems approach are considered in more detail in a recent paper (1).

Earlier university-based projects involving centralized support for experimental instrumentation have been reported at the University of Oregon (C. E. Klopfenstein), the University of Nebraska (C. L. Wilkins) (4), and at Purdue University (R. B. Harrington) (2). Current projects also exist at the University of Kansas (P. W. Gilles) and at the University of California, San Diego (K. Wilson). The systems named are generally departmentally oriented, the discipline in four out of the five cases being chemistry. An interesting project recently started, whose objectives are somewhat more general, is at the Andrus Gerontology Center of the University of Southern California (G. Reese). Here it is the intent to support a variety of minicomputer applications in the Center by a system very similar in outline to MISS, although with more uniformity as to types of minicomputers supported. One particular commercial system of this type has also been announced, the Hewlett-Packard Distributed Systems (3).

Applications of such systems concepts to clinical care or similar ongoing operations seems to be in a rudimentary stage of development, although in a few cases systems designed for instrumentation support may be evolving toward more general on-line applications.

Compared to most of the other developments that have been reported, MISS has more of an orientation toward the idea of a general system, flexible as to interaction with users and their minicomputers, and with centralized facilities. Thus little assumption about the particular uses of the minicomputers is made, only that it is desirable to have the ability to access enhanced facilities on a continuing basis. The enhanced facilities back up "operation mode" for the minicomputers, involving data collection, automatic control and human mediated control, and provide for ancillary analysis and follow-up activity. At least as important, however, is the backing up of "development mode," where the facilities essentially provide an expanded configuration for program development purposes.

MISS ORGANIZATION

The Minicomputer Interfacing Support System is basically conceived as a hierarchical hardware/software configuration for support of on-line minicomputers. At the lowest level of the hierarchy are the minicomputers, of a variety of types and dedicated to laboratory and other on-line functions. It is assumed that they are to be used a good deal of the time in "standalone" fashion, and are controlled in this operation by the user or department that has acquired them. Because there are occasional needs for enhanced peripheral and other computing services, however, these are equipped with an interface so that they can be "hooked in" to the intermediate level of the MISS system, to a dedicated system called the Minicomputer Operation Monitor (MOM). This level provides, on a shared basis for all minicomputer users: *1*) extended input/output services (card reader, printer, and the like); *2*) mass storage services (mainly disk, but also magnetic tape); and *3*) access to the highest level of the system, where are maintained a set of specialized support services under the name of Data & Algorithm Distributor (DAD), and also a general system connection to central computing facilities. The intermediate level system is implemented on a PDP-11/45 computer configuration, and the highest level is at present the 370/168 system run by the University Computation Center. In the larger concept of MISS, however, the third level is a network of major computer systems. This *a*) allows a much broader class of backup services to be provided, and *b*) permits replication of MISS at any desired node of the network by simply duplicating the 11/45 configuration.

The three-level hierarchical organization of MISS is essential to understanding its conceptual basis. The minicomputers at level 0 of the hierarchy are served by extended storage and input/output facilities at level 1 of the hierarchy, and extended computing facilities at level 2.

The level 1 facilities are shared among MISS users only, and it is to these the designation MOM applies. The level 1 hardware configuration, the DEC PDP-11/45, has peripheral resources that may be shared. In particular, the magnetic tape units, paper tape reader/punch card reader and line printer may be thought of as extended I/O units for the minicomputers. The console terminal and the disk units function more as system resources for MOM itself, although users can obtain limited disk space allocations as part of MOM services.

The level 2 facilities are general purpose and shared with others not part of MISS. For the present implementation, as mentioned, these facilities are the

University Computation Center IBM 370/168 system. The services supplied at this level, however, are distinguished as those particularly developed and maintained for MISS users, the DAD services, and those available to any user of the Computation Center. Users of MISS have access to the 370/168 in both these modes. Similar considerations would apply to a more general network at level 2.

The motivation for a three-level system of this form, instead of a two-level system where minicomputers are connected directly to general-purpose computing facilities, is to provide flexibility for the minicomputer user and also the advantages afforded by having a minicomputer interfaced primarily to a dedicated system, serving a group of users with similar objectives.

REFERENCES

1. Ashenhurst, R. L. Hierarchical Computing. In: *Networks for Research and Education*, the report of the EDUCOM-NSF General Working Seminars on a National Science Computer Network. Cambridge, MA: MIT Press, 1974.
2. Beckwith, D. Computerization—A planned expansion. *Am. Lab.* 5: 77–83, 1973.
3. Hewlett-Packard. *9701 A/B Distributed Systems* (sales brochure), 1973.
4. Klopfenstein, C. E., and C. L. Wilkins Multicomputer processing in laboratory automation. *Proc. 1971 Fall Joint Comput. Conf.* 1971, pp. 435–40.

Session V—Networks and shared facilities

Rapporteur's summary

KENT R. WILSON
University of California, San Diego
La Jolla, California 92037

Our topic is the sharing of computer resources by intercommunication. I'll summarize the session. in four parts: what's been done, why do it (the advantages), why not to do it (the problems) and what needs to be done in the future.

WHAT'S BEEN DONE

There are three basic configurations of computer networking: *i*) a single computer shared by many inputs and outputs; *ii*) a set of computers linked in a horizontal network, with a parallel distribution of functions; and *iii*) a set of computers linked in a vertical network, with a serial hierarchy of functions.

In size, networks range from those in a single laboratory to the intercontinental, and now include the alphabet soup of systems described by E. Bowdon. Specific examples of each of the above types are: *i*) PROPHET, a nationwide, discipline oriented single computer network described by W. Raub; *ii*) ARPANET, a largely horizontal, intercontinental network of several dozen computers described by F. Heart; and *iii*) several hierarchical systems from the local to national scale, largely vertical, described by R. Harrington, R. Ashenhurst and E. Meyer. Applications to the life sciences range from crystals to cornfields and include X-ray crystallography, protein and nucleic acid structure and function, pharmacology and toxicology, environmental science, and agriculture.

WHY—ADVANTAGES

The goal of sharing and networking is to achieve the best use of resources; to provide more computing resource with more convenience to more people for less money.

Computer needs are increasing faster than computer resources; therefore we need to use these resources more efficiently. In particular, we need to help *people* be more efficient, as they are the most expensive part of problem solving with computers. There is a growing realization that no computer can be all things to all men, and thus that new modes of computer usage are called for.

> *Sharing of a single computer system* makes practical the implementation of specialized computing systems. The advantage is tuning to the task the hardware, software, data bases, and consultants to maximize problem-solving ability in a particular area of discipline. If the focus for a computing system is narrowed in order to deepen its usefulness, its geographical outreach must be broadened to gain a large enough user group: therefore networking. Example areas from this session include pharmacology, toxicology and crystallography.
>
> *Horizontal networking* provides redundancy and thus reliability; a larger user

group and thus load leveling. Computing resources available to the user can be enriched by the possibility of accessing a variety of specialized machines allowing, for example, programs to be run untranslated on the type of machine for which they were coded and the setting up of specialized large data storage and manipulation facilities. Such networks naturally encourage the sharing of data and programs, the development of standards and protocols, and the sprouting of competition among alternative suppliers of services, leading to improvement of service. More subtly, human–human interaction is fostered; if we link computers we also link the people using them, aiding the development of discipline-oriented communities.

Vertical networking allows the linking of the convenience and cheapness of small dedicated computers with the power and peripherals of larger computers. Small inexpensive computers can be close-coupled to experiments or persons to provide fast dedicated response capability (remembering that distance alone may make fast response from a remote computer impossible). A larger more expensive computer may be shared in a slower response mode by many close-coupled smaller computers, providing computational and peripheral backup when needed.

WHY NOT—PROBLEMS

The major limitations to the spread of networking are not in network technology, but are political, organizational and economic.

Political. Why should a group having control of a dedicated resource work hard to make it more available to others and thus less available to themselves? Why should a campus computer center director encourage his captive clientel to spend their money elsewhere through making networking available?

Organizational. How can channels be implemented so that a computer center responds to the needs of its remote networked users?

Economic. Why should anyone be among the first to join a network, before there is anyone else interesting to talk to? Are present types of networks economically viable if subsidies are removed?

It is clear that networks themselves are not a solution, but only provide an opportunity for solutions to be found.

WHAT NEEDS TO BE DONE

Networking presents an outstanding opportunity for broadening the access to and raising the level of computer support to the life sciences. But shared computing needs shared effort; it can't be done by a single researcher. Networking is a reality which must be dealt with. User practices, institutional procedures and governmental policies all need to adapt if the advantages of networking are to be gained.

Active pushes are needed in the following areas:

Making networking available to smaller, less sophisticated users, which will require hand-holding. Now, too often, the rich get richer; those who are already well endowed with computer access get more through networking.

Overcoming of political, organizational, and economic impediments.

Encouragement and subsidy of protocol development and standardization of access, even across different networks.

Bringing specialized and discipline-oriented computing centers into sophisticated networking, for example the Health Sciences Computing Facility, the National Center for Atmospheric Research, Lawrence Berkeley Laboratory, Brookhaven National Laboratory, and so on. In addition, serious consideration should be given to the implementation of new discipline-oriented centers now made practical by networking.

The drive behind these pushes must come from potential user communities, for example in the life sciences, who find that by sharing they gain more than they lose.

REFERENCES

1. Greenberger, M., J. Aronofsky, J. L. McKenney and W. S. Massy. *Networks for Research and Education.* Cambridge, MA: MIT Press, March 1974.

Session VI

Meeting the costs

Computing and competing for research dollars

RUTH M. DAVIS, *Chairperson*
National Bureau of Standards
Gaithersburg, Maryland 20760

The objective of this session is to provide you with a sampling of the concerns and realities of practical computing in life science research. Accordingly, we want to: *1*) address the problem of computer costs and project how these costs could be reduced via aggregation of the life science market; *2*) survey and assess Federal support for computing applications; *3*) consider means of furthering the practical applications of research by expediting communications between the medical research community and the community of practicing physicians; and *4*) look into the political climate of funding life science research and examine the specific appeal of computer applications.

To do this, we have on our panel today, well-known experts to discuss these questions from the viewpoint of industry, the viewpoint of a Federal research policy maker, the viewpoint of a practicing physician, and the viewpoint of Congress.

Meeting the real costs of the computer as a research tool in the life sciences

RUTH M. DAVIS
National Bureau of Standards
Gaithersburg, Maryland 20760

THE FOLLY OF DISASSOCIATING MEDICAL RESEARCH FROM ECONOMIC CONSIDERATIONS

Medical research and economics are strange bedfellows. It is rare indeed to find medical researchers introducing economic considerations into their projects. It is also quite unusual to find economists voluntarily delving into the real costs of research.

What does bring economics—or more specifically, costs—and medical research together? Usually the binding agents are the scarcity of available resources coupled with requirements for setting up priorities on the basis of which the scarce dollars are allocated. Further, one can always be sure that the requirement for establishing priorities is always imposed by groups outside the medical research community.

Why are research and economics so seemingly incompatible? Probably, the real culprit is the honest confusion that researchers use to their own advantage. How often have we all been told that research is serendipitous and that you cannot put a price on creativity? And that is certainly true some of the time.

But, there are many kinds of research. Activities tagged "research" by some researchers are called novel applications by others. Some "novel applications" may be just simple repetition in a new geographical location.

Students of science and technology have found it convenient to distinguish between: *invention* (or basic research), *innovation:* the first successful applications of invention (or basic research), and *diffusion:* the widespread repetition of innovation. Most of the projects that have been discussed during this conference seem to fall into the innovation category. To say that innovation cannot be costed is sheer nonsense or technological laziness. Only invention defies costing or economic fitting. And yet, very few economic analyses generally accompany innovation or diffusion in the medical community.

COMPUTERS AND COSTS

Throughout this conference the applications of the computer as a research tool have been discussed. I am sure you have been impressed with the computer's ability to simulate analogs to living systems, monitor experiments, manipulate extremely large sets of research data, and perform countless research chores.

However, we know that the computer is still a research tool of only a privileged lot of life scientist researchers today. As you consider the computer's potential in your research, you may be curious about costs. Certainly, policy makers, Congressmen, economists and research sponsors should be curious about the economics of computers as research tools. This session is directed towards discussing the costs and real world context of computing.

COMPUTER COST JUSTIFICATION—NECESSARY OR NEEDLESS?

Accordingly, I initiated a research for the modus operandi of justifying computer costs in life science research.

After querying authorities in government agencies from NIH through NASA's Technology Utilization Office, interviewing knowledgeable people at the National Academy of Sciences and their Institute of Medicine, tracking appropriate advisors to the health subcommittees in Congress, talking with specialists at the Library of Congress, questioning knowledgeable technical press people, and poring over print-outs from three computer literature searches (NTIS, MEDLINE and SSIE), my staff became suspicious that I had sent them on a snipe hunt.

Computers are generally not economically justified in analyses and review of basic health research. Cost/benefit studies of their merit are generally not used to allocate research funds nor to place computers in Federal laboratories. In fact, my staff reported that the concept of justifying the use of computers on a dollar basis seemed foreign in the life science community.

"Do you consider the cost/benefits of computers used in research?" my staff asked authorities at NIH.

"Talking in terms of economic justification of computers in research is knuckling under to the GSA and OMB conception that everything can be costed out. Research CANNOT be costed out!" one NIH official protested. Another collaborated this sentiment: "You cannot come up with a cost/benefit analysis of any aspect of basic research. There is no way of assessing and assigning numbers to research progress."

From the Institute of Medicine at NAS, my staff learned: "Research contracts are awarded mostly by interest in a particular area. Rejection of a contract is usually on a substantive rather than an economic basis—research is serendipitous."

A biomedical specialist at the Library of Congress responded to our inquiry with: "I wish you were a Congressman because I'd like to sink my teeth into that project." He lamented that although he had seen references to computer applications in life science research, the Library has never been called on to research the general subject for Congress, much less the subset of economically justifying computerized research.

FUNDING COMPUTERS FOR LIFE SCIENCE RESEARCH

Congressional interest in funds spent specifically for computer applications in health research was minimal. As an advisory to the Senate Health Subcommittee put it: "We are not concerned with how research is conducted or its cost/effectiveness. We just decide what to fund and leave the spending decisions up to the agencies."

The agencies performing or sponsoring research reported that the decision to fund a research project would not depend on computer costs. An NIH authority summed it up this way: "In awarding research grants, we review applications primarily for scientific merit. We do not dictate how the dollars will be used."

The costs of research tools are labeled a minor consideration by some life science authorities who assert that funding is based on whether a project: is do-able; is desirable to do; and will be done properly, based on the track record of the researcher and the elegance of his plan.

Hereby it would seem that the costs of a particular project, whether or not it is inflated by computer costs, are not a major factor nowadays in granting basic research funds—within bounds, of course, of what is reasonable. This expression of unconcern for computing costs appears to be a superficial interpretation that is hopefully fallacious. Indeed, some authorities are probably quite concerned about research costs generally. However, this concern for costs needs to be clearly manifested at the policy level.

COMPUTERS IN LIFE SCIENCE RESEARCH LABS

It is important to recognize that computers are being placed in more and more life science research laboratories. As one NIH manager expressed the situation: "Our scarcest resource is intelligent manpower and the computer is, therefore, an invaluable productivity increaser." Herein, we may view the computer as a sort of "Hamburger Helper" that makes the mental meat go farther.

One technological factor making more fluid the decision-making that underlies the acquisition of computers for research has been the amazing proliferation of minicomputers.

Minicomputers, admirably suited for many laboratory and some other research projects, now cost in the $5,000 to $20,000 range. It is only when research or experiments demand large-scale data manipulation, heavy computation, or massive programs that minicomputers are not adequate. With the cost of minicomputers still decreasing and with their capability for real-time measurement and sensor input still increasing, it may well be that their cost should play a minor role in research fund justification. Certainly, it is significant that in 1973 over half the U.S. computer inventory of 135,000 computers were minicomputers. Five years earlier, there were hardly any minicomputers in our computer inventory.

The decision to put a dedicated minicomputer (frequently tied into a network) in particular laboratories is usually based on the judgment of the professional, scientific leadership. It is not commonly an administrative decision. Computers are placed where they are integral to the productivity of the researchers, provided they are doing good work and deserve further investment.

Although the double criteria of "needing a computer" and "being deserving" could lead to some difficult judgment calls when one lab is more deserving but needs a computer less, this conflict is resolved via the leadership's subjective judgment. Cost/benefit analysis in dollar terms seems not to be a felt need of researchers or research sponsors.

IS THE COST OF THE COMPUTER AS A RESEARCH TOOL RELEVANT?

The research world is not now demanding cost/benefit justifications of computers as research tools in the life sciences. Nevertheless, the question of whether computer costs are relevant to you in planning your research is one we hope to address in this session.

WHEN SHOULD COMPUTERS BE USED?

Currently, computer use in life sciences research ranges from requisite to ridiculous. Some research could not in fact be done without computers. For instance, one computer critic admitted that his own research to determine how nerves conduct impulses depended on a computer. His experiment involved sending electrical impulses through frog nerves. The computer collected thousands of data points within the 1–2 sec it took the nerve to relay the impulse, and amassed this data for several thousand sequences.

But this critic complained that computers occasionally become a distracting toy. He disparaged scientists who make computer games out of their research, "motivated by the desire to produce pretty printouts and stymied from sophisticated thinking."

Herein, we see that the computer's contribution to—or distraction from—research productivity is relevant to its desirability, although cost effectiveness per se is presently distasteful.

THE SCARCITY OF FUNDS

No one now questions that life science research is being done in an era of fund scarcity. Resources have shrunk and the effects are beginning to be felt.

Hard times often require reassessment of goals, methods and priorities. The decreasing availability of resources may lead to modifications in the mode of funding research.

Perhaps the favorable trade-off between research benefits and costs of using computers in a particular research project will become a recognized selling point in the competition for scarce funds. The computer should thrive where it is integral to efficiently accomplishing the research mission; the computer should become extinct as a luxurious convenience, toy and showpiece.

Perhaps, we have already then come to grips with a primitive schema for the economical and practical use of the computer as a research tool in the life sciences.

REFERENCES

1. ANON. *Fiscal year 1974 Annual Report.* Biotechnology Resources Branch, Division of Research Resources, NIH, Bethesda, MD., 1974.
2. ANON, *NIH Public Advisory Groups.* DHEW Publication No. (NIH) 74-12. NIH Committee Management Staff, Bethesda, MD., 1974.
3. ANON. A policy for biomedical research. *J. Med. Ed.* 46: 691–743, 1971.
4. CARTER, G. M. The effect of some NIH policies on biomedical research in medical schools. Rand Corporation, Santa Monica, CA. In press.
5. EAVES, G. N., D. B. RIFKIN, H. H. GEE, T. E. MALONE, R. ROSS AND R. T. SCHIMKE. The project-grant application of the National Institutes of Health. *Federation Proc.* 32: 1541–1550, 1973.
6. GEE, H. H. Organization and influence in biomedical computing. *Rev. Inform. Med.* 1: 281–289, 1970.

EDITOR'S NOTE

The preceding paper raises sharply and provocatively the problems of evaluating research on the $cost–$benefit formula familiar to economic analysts, with special reference to the research-grant funding policies of governmental agencies. It seems, then, highly relevant to review the procedures and policies actually employed by granting agencies in assigning priorities to research applications, especially those in which computers are involved.

For many years the Federal Government has relied heavily on the peer review system for recommendation of public-fund research expenditures. At the NIH, incoming research grant applications are assigned to the appropriate Institute, and sent for review to one of about 50 Study Sections whose members are expert in the field of research involved. Each Study Section is made up of roughly a dozen eminent nongovernment scientists, who serve 4-year terms; these groups carefully evaluate each application, determine whether the application should be approved or disapproved based on scientific merit, and assign to each approved application a priority ranging from 100 (top) to 500 (bottom). This priority constitutes a recommendation for funding to the next review body, a National Council, and is weighted very heavily in making the decision whether or not to fund the application from the scarce research dollars available. The various NIH Study Sections and their membership are listed in (2), published annually. The entire NIH review process is well described in (6), with particular reference to computers. Informed and very helpful advice to the grant-application writer is given in (5).

Grant applications with a heavy computer component are, at the NIH, usually referred to the Study Section on Computer and Biomathematical Research, a cross-disciplinary Study Section comprised of physicians, other life scientists, computer scientists and mathematicians. This group has been greatly concerned for well over a decade with the cost-effective use of computers in the life sciences. The *cost, potential benefit* and *risk of failure* of the applications are evaluated on a comparative basis, as well as the necessity, appropriateness and cost of computer equipment involved, and priorities assigned accordingly. Computer Study Section members frequently perform this function for other Study Sections as ad hoc reviewers, so that the expertise of its members is widely available. A similar policy of comparative evaluation of cost, benefit and risk is followed by the Division of Computer Research at the National Science Foundation. This type of analysis, the comparative evaluation of benefit, risk and cost, seems at present the most tenable basis for allocation of limited research funds.

Clearly, the benefit and risk elements are the hardest to evaluate. (How does one estimate the dollar value of uncovering the structure of DNA?) For basic research, this simply cannot be done; the benefits are not costable. However, a *comparative* analysis of competing applications in risk and benefit terms can be done, and is done by most granting agencies. As the thrust of the application moves from invention through innovation to diffusion, it becomes less difficult to attach a dollar value to benefit, and the uncertainty in evaluation of risk decreases. To help reduce the uncertainties in risk evaluation, most granting agencies advise the applicant to submit evidence of substantial pilot work.

While the "cost" element of the cost-benefit-risk function is the most easily evaluated in dollar terms, it too can be a bit elusive. In an era of ever-increasing costs of almost everything, computer costs (especially of minis and midis) are declining, in some cases dramatically. For a grant application which promises a beneficial and readily-diffusible product, one must evaluate the costs not solely on the basis of today, but perhaps on what costs will be a few years from now. For many applications, costs must be looked at in a time-dependent and dynamic context. While this complicates the picture, reasonable extrapolation of costs can and should be made whenever necessary.

Unfortunately, many persons are not aware that this type of comparative cost-benefit-risk analysis is going to be applied by most responsible granting agencies, and the individual scientist who ignores this does so at the peril of having his grant disapproved, or his budget drastically slashed. It is obvious from the preceding paper, and from many grant applications which this editor has reviewed, that the NIH has often failed to communicate, even to some in government, what the real working policies of grant review actually are at the nuts-and-bolts level. The NIH system is an excellent one for assuring that federal research dollars are spent in the most cost-effective way possible, and should constitute a model for funding agencies; if it is to be so, accurate and complete descriptions of the actual review process should be widely circulated.

In fact, the problems of costs and benefits of research have already received intensive study at the NIH policy-making level. An important study was carried out by a special committee at NIH initiative (3). Based in part on specific studies of the costs of death and illness, and benefits reasonably to be expected from medical research, this report estimates the annual return to the government in income and excise taxes as a result of improvements in health care to exceed the entire NIH appropriation. Specific policies are suggested for funding grant applications, and for setting the general funding level for biomedical research. Special sections are devoted to cost-benefit analysis and to funding criteria, although these are broad in outline and general in nature. An up-to-date overview of NIH funding policies is given in (4). This report includes objective measures of research output and quality, which in turn are used to survey critically present funding practices. Also included is a novel statistical evaluation of Study Section performance, including effectiveness, flexibility and the like.

More specific policies regarding computer funding and cost-effectiveness are given in (1), prepared by the NIH Biotechnology Resource Branch. (The NIH Research Resource program has been heavily responsible for the diffusion of computer techniques in the life sciences, and for the development of new computer techniques especially for biomedical research.) This report contains metrics not only for cost effectiveness, but also for the diffusion of computer technology at particular life science computer facilities.

Dr. Davis calls for clearly-stated policies on cost-benefit analysis in research; it would seem, for the NIH and NSF, that stated policies which reflect the actual procedures and criteria now used would serve this purpose. In turn, other granting agencies should become increasingly aware of the importance of cost-benefit-risk analysis, and should employ such comparative analysis with increasing sophistication and rigor.

William Siler

Industrial view of computer applications in the life sciences[1]

JORDAN J. BARUCH
Dartmouth College
Hanover, New Hampshire 03755

The phenomenal growth of the use of digital computation in such fields as airline reservations, hotel-chain record keeping, banks and credit agencies overshadows its much slower growth in such areas as biomedical research, retail stores and architecture. To the researcher, to the government agencies that sponsor biomedical research and to the layman, it seems inequitable that the benefits of computation flow so easily to ease the pain of the air traveler but still contribute so little to the life sciences.

It would be easy to blame the greed of industry or the parsimony of granting agencies for this sorry state—easy but probably invalid and certainly unproductive. If we stipulate that our goal here is to change this trend rather than engage in political polemics, we can dispense with the urge to blame, search for reasons for the trend, and discuss policies and strategies that may eliminate those reasons.

Computing is a subset of the general class of phenomena called technological change. Thus we can draw some understanding of its progress from the research that has been done in the field of technological-change diffusion. While much of that research deals with sociological parameters, and while the sociology of the biomedical field certainly contributes to its reaction to computation, those parameters are considerably less malleable, less subject to short-range change than are the equivalent economic parameters. Since the aim of this paper is to suggest policies and strategies that can, in the near future, increase the rate of diffusion of computation in biomedicine, we shall restrict our analysis to two economic phenomena: *1*) nonappropriability of benefits, and *2*) market disaggregation. The first of these seems predominant in the health care delivery sector of biomedicine, the second in the research sector. To demonstrate how, let us examine some earlier research findings on technological change in industry.

Mansfield[2] has shown that the rate of diffusion of an innovation among organizations in a common industry is linearly correlated positively to a measure of the perceived profitability of that innovation relative to computing investment opportunities (P), and negatively to the portion of the organization's total assets that are

---

[1] This paper is based on work performed while the author was Lecturer in Business Administration at the Harvard Graduate School of Business Administration and was supported under grant number 7943-2 from the National Science Foundation.

[2] MANSFIELD, E. *Industrial Research and Technological Innovation*. New York: Norton, 1968, chapt. VII.

put at risk (S) in investing in the innovation. Specifically, where R is the diffusion rate of the innovation in % of the firms in the industry adopting per year,[3]

$$R = -K + 10.6P - 0.54S$$

where K is industry dependent and takes on values from 6 to 12 in the twelve industries investigated.

Hazardous as it may be to extend these findings to the generally nonprofit area of health care delivery, we shall do just that albeit in a more quantitative form. In the biomedical field, we can consider two diffusion rates, R_S and R_A. R_A is the rate that results from the advantage and risk perceived by the health-care or research administrator and R_S is the rate that would result if the administrator had society's perception of advantage (P_S) and risk (S_S). Our problem is that $R_A << R_S$. Let us examine the health-care delivery area first. For low-cost innovations or large hospitals, the last term, S_A, becomes small and we may focus on the difference between P_A and P_S, a difference the economist terms a market externality.

A primary difference (although not the only one) occurs when the advantage considered in P is cost reduction. As a first approximation, P_S can be taken as the ratio of the annual cost saving per dollar of investment required to the annual cost of debt service and amortization over the life of the innovation. For the Federal Government, this denominator ranges from about 0.1 for an infinite-life investment to 0.25 for a 5-year investment. Thus, for any long-term investment such as computation that saves over 10 cents per dollar invested, the value of P_S will exceed unity.

For the individual administrator, however, such is far from the case. Under modern reimbursement practices, any cost savings produced through innovation do not flow to the adopting organization. At audit time, reimbursements are rapidly reduced. Thus, if all innovations were bought on borrowed money, if the interest on that money and the depreciation of the equipment were allowed as an expense and if the loan direction equaled the equipment life, P_A could never exceed unity regardless of the cost saving!

If we use an estimate of \$0.50 saving per \$1.00 invested and a loan cost of 10% to the hospital, with a 5-year equipment life, we can estimate the ratio of R_S to R_A. With no a priori estimate of K, we'll assign it the midrange value of 9 and set S equal to zero.

$$R_S = -9 + \left(\frac{0.50}{0.25}\right) 10.6 = 12.2$$

$$R_A = -9 + \left(\frac{0.25}{0.25}\right) 10.6 = 1.6$$

$$R_S/R_A = 7.65$$

Thus society would expect 12.2% of the hospitals to adopt such an innovation per year, while only 1.6% would actually do so. Society's expected rate would thus be over seven times as great as that actually experienced! While these figures represent the grossest of approximations, the message is clear.

To overcome this problem, either society can pay for such innovations directly, or some method must be found to bring the perceived interest of the administrator more in line with society's. An example of the latter policy might be as follows:

1) Based on level of care, quality, or some other output measure, have the AHA or a similar body classify all hospitals into N classes. Such classification is to be based on examination of medical performance, is to be independent of any costs or prices, and is to be periodically reviewed.
2) Compensation to a hospital in any class is to be based on the average of the costs incurred by all hospitals in that class.
3) Government guaranteed loans are to be made available to hospitals for capital investment.

[3] This expression is somewhat different from Mansfield in that it expresses the diffusion rate over the period from 20% adoption to 80% adoption.

Basing reimbursement on class-average cost results in cost savings accruing largely to the organization making the innovation. Government guaranteed loans set the hospital discount rate near that of society (and, with appropriate provisions, equalize S_A across hospitals or set it at zero). Performance stratification still permits as many degrees of choice in level of care as desired, and the review process prevents cost saving at the expense of quality.

In the research area, as opposed to the medical care area, another force seems at work. To get at that force, let us examine the components of P_A from the researcher's viewpoint. To do so, we remember that P is the ratio of the perceived gain per dollar invested in the innovation divided by the perceived gain per dollar invested in alternate investments. Thus, if the perceived gain from computation is artificially reduced or if the cost of computation is artificially elevated relative to other investment opportunities, the diffusion rate of computation will be inhibited.

In the case of biomedical research, both of these phenomena seem to be produced by the structure and character of the biomedical research market. The perceived benefit (however that benefit may be measured) is generally taken to be some "true value" of benefit, times the a priori probability the decision maker assigns to securing that benefit. Much of the sales activity of reputable manufacturers goes into raising that assigned a priori probability. Sales calls that "spread the work," demonstrations, trial periods, and the like are all aimed at that goal. In the highly specialized and diversified market of biomedical research, such probability-raising activities must be tailored to many different kinds of buyers from crystallographers to geneticists. The rational seller, therefore, chooses to spend his sales dollars on more aggregated markets where one body of information can sway many prospective purchasers and biomedical research becomes an "undersold" market because of its disaggregation and diversity.

A similar problem seems to exist in the area of relative cost, the denominator in P_A. The effective cost of computation, to the researcher, is comprised of the direct cost and a group of such start-up costs as the programming costs, the educational costs, and so on. In the case of the airlines, those costs may be spread over many purchasers either by the manufacturers working to agreed-on specifications or by consortia of users who undertake the start-up activities. Both techniques have been used. In the case of biomedical researchers, disaggregation again impedes diffusion, this time through elevating investment cost relative to other opportunities lacking such start-up costs. No standards exist and no combination of researchers provides for the needed economy of scale such as are found in airline reservations.

The Prophet program under the NIH is a first-order attempt to compensate for both the cost elevation and the a priori perceived benefit reduction by providing a well publicized, standardized system for pharmacological researchers. Here the start-up costs and uncertainty reduction activities have both been undertaken directly by a government agency with a well conceived and managed program. While Prophet bodes well for pharmocological investigations, and while much of its programming is widely transferrable, it is unlikely that many specialities will find a skilled champion in government service to pursue a similar route.

Such a route, however, is only one possibility. Any group of specialists, particularly those bound together in a professional association, can serve a similar function. Establishing a system that is standardized and specified out to some operating boundary and publicizing the same will have several effects. First, for smaller manufacturers whose market would be significantly enlarged by sales to a biomedical segment, the standardized system offers a marked attraction. His small sales force need only be familiar with that speciality, a network of prospects exists, and features extraneous to that system can be eliminated (or essential ones added) with significant cost control implications. For the group of specialists, significant advantages also exist. The activity can, for example, be organized under a professional society aegis. That operating group can then solicit support directly from outside agencies as well as from the members, to design and publicize the system. By improving the a priori perceived probability of securing benefits and by reducing or essentially eliminating the start-up costs, one would expect such an aggregating strategy to produce the same diffusion rate for computation in biomedical research as has been experienced by other capital investments.

Our sample policy in health-care delivery and our sample strategy in biomedical research are, of course, just that—samples. Many other routes to increasing R, the rate of diffusion of computation, can be conceived and should be struggled with. Salient to all, however, will be an understanding of the determinants of technological diffusion rate and the consequent recognition of the economic barriers that now inhibit it. It is our belief that the intelligent use of benefit-appropriability policy and of aggregating strategies can markedly improve the diffusion rate for computation in the field of biomedicine.

Sponsorship of computing research in medicine and biology

HOWARD MORAFF
National Science Foundation
Washington, D. C. 20550

When I set out to write this presentation, I had an idea that it would be an easy job: I'd just call all my friends in Washington and ask them about support for the use of computers in life science applications. Well, it wasn't that easy. That support is scattered, and most of it is imbedded in support of research in the disciplines. The survey I promised is really more of a sampling, and is likely not complete. The best advice I can give to those of you who may wish to seek support is to follow up these leads by letter or by phone, and ask each time for suggestions on other interested agencies.

Federal support for life science activities involving computing is obtainable from the Department of Health, Education and Welfare (HEW), the National Science Foundation (NSF), and others.

The Department of Health, Education and Welfare (HEW) is the heaviest supporter of computer applications in the life sciences, with several direct and a number of indirect sources. The Biotechnology Resources Branch (BRB), Division of Research Resources, NIH, under Dr. William Raub, supports establishment of resources for computing as well as other high-technology instrumentation applications. Grants are also made for resource-related research projects.

The BRB has instituted a Program for Shared Biotechnology Resources. An announcement dated May 10, 1974 is available. The intent is to make existing resources available to qualified biomedical scientists in the same and other institutions and to facilitate the sharing of unique facilities and capabilities across institutional boundaries.

The National Institute of General Medical Science (NIGMS) has a Bioengineering Program under Dr. James Dickson, which funds projects in biomedical research, including computer applications. HEW has also supported computing resources through the Health Care Information Systems and Technology Division, Bureau of Health Services Research, Health Resources Administration, under Dr. Bruce Waxman, and the Special Program Staff, Bureau of Health Resources Development, Health Resources Administration, under Dr. Katherine Johnson.

Support for a variety of research applications of computers is also imbedded in grants for research in the disciplines supported by the categorical institutes of the NIH; many of these applications, especially those with heavy computer and mathematical components, are reviewed by the Study Section on Computer and Biomathematical Research. The Executive Secretary of this Study Section is Dr. Bernice Lipkin. Support for educational applications can presumably be obtained from the National Institute for Education.

The Veterans Administration (VA) is heavily involved in applications of computers in health care and the medical sciences, but must be treated as a special case. The VA is not a granting agency: all the work is essentially done "in-house" at the 170-odd VA Hospitals, in some cases in cooperation or collaboration with neighboring medical schools and hospitals. The system is structured to enable easy sharing of these vast resources between the VA and the medical and medical research communities, with very little dollar exchange involved. Support for computing activities by other Federal agencies (such as the Advanced Research Projects Agency of the Department of Defense, and the Office of Naval Research) is essentially imbedded in grants and contracts in the various research areas, with a couple of isolated exceptions.

The National Science Foundation is a primary source of support for basic research. Applied research is also funded, in the context of meeting national needs, and some large national centers (for radio astronomy, atmospheric research, and so on) receive support from the NSF. Computer application support is imbedded in many of these activities. The Technological Innovation in Education Group of the Office of Experimental Projects and Programs, Education Directorate, NSF, supports research in educational uses of technology, including computing applications. Erik McWilliams is the Acting Director of that group. An example of the activities supported by that group is the PLATO computer-assisted instruction system at the University of Illinois. The Division of Computer Research of the NSF supports both basic research in computer science and efforts to find applications for computer science concepts and techniques. This last is the essential thrust of my own Techniques and Systems Program, in the Computer Applications in Research Section, Division of Computer Research, NSF. The program makes grants to researchers in computer science and in the traditional research disciplines, to help find new ways to use computers in the conduct of research. The emphasis is on innovation, either in a global sense (i.e., no one, anywhere, has ever done this particular thing) or in a local one (i.e., this technique was developed for a particular research application, but with some ingenuity, we can transfer the techniques to other application areas). The grants are mostly small, supporting a person or two for a year or two, in pursuit of a specified advance in computer-based methodology (although larger grants are also made which support efforts having larger potential impact and the need for a concerted effort by a substantial group of investigators). Some areas of exploration have a potential use which spans several or many disciplines. Techniques and applications of computer graphics, on-line real-time signal acquisition and processing, and various aspects of human–machine interaction are examples of broad applicability: these have received heavy emphasis.

To get more information about support prospects, you can ask the various Federal agencies for their program announcements. Many also offer annual reports, summaries of recent awards and other descriptive materials, either free or at nominal cost.

In conclusion, Federal support for the acquisition and use of computers in the life sciences is available from many of the agencies that normally fund research and educational activities in these disciplines. Applications for funds to acquire computer systems and resources normally are judged on the scientific merit of the research to be conducted at the resource. Such applications may be enhanced by combining the interest, needs and participation of a number of investigators, thereby taking advantage of the economies of scale. Research and development of highly specialized and unusual resources may be considered for support by the Biotechnology Resources Branch of the NIH, and explorations of new computer methodology for research and educational applications are of interest, respectively, to the Techniques and Systems Program and the Technological Innovations in Education Group of the NSF. Both the staff of BRB and that of the Techniques and Systems Program (i.e., this author) hereby volunteer to help you find the appropriate programs for your computer applications.

REFERENCES

1. *Guide to Programs, Fiscal Year 1974.* National Science Foundation NSF74-7.
2. *National Science Foundation Annual Report, 1973.* NSF74-1, $2.35 from Superintendent of Documents, U. S. Government Printing Office, Washington, D. C. 20402.
3. *National Science Foundation Grants and Awards 1973.* NSF74-2. $3.30 from Superintendent of Documents, U. S. Government Printing Office, Washington, D. C. 20402.
4. *Program Description, Biotechnology Resources Branch, NIH, October 1973.* (also booklets *Modes of Resource Support and How to Apply for Grants,* October 1973).
5. RAUB, W. F. Biotechnology cranks out new research hardware. *Lab. Management* January 1973.
6. *Summary of Awards, Fiscal Year 1973.* Division of Computer Research, National Science Foundation. NSF74-10.
7. *Summary of Grants and Contracts Active in June 30, 1973.* National Center for Health Services Research and Development (now, Division of Health Care Information Systems and Technology), Health Resources Administration, U. S. Department of Health, Education and Welfare, Rockville, Maryland 20852.

A medical practitioner's view on computer applications in the life sciences

RALPH CHRISTENSEN
Boise, Idaho 83707

I have been sitting here the past 3 days listening to you people talk to each other. I have not understood much of what you said, and I have had a song running thru my mind which asks "What Are You Doing In My World?" I think we could ask that question of each other. I know that many of my colleagues wonder what the biomedical research people are doing in the practice of medicine. I think many of them look with some anxiety and fear at what they think you are doing.

I think I have been fortunate in being able to practice in a community where the attitude was one of careful examination of the quality of care that was being given in local hospitals. I hear a lot being said now about medical audit. Medical audit has already been practiced for at least the past 25 years at Sacred Heart Hospital in Eugene, Oregon. In fact, I believe that we have probably been more effective than the Professional Standards Review Organization, which the American Medical Association enthusiastically endorsed just yesterday, will be.

We are all aware of the changes in the practice of medicine that have taken place since the Second World War: the diminishing number of solo practitioners; the increase in group practice; the marked changes in funding which occurred with the legislation in 1965; and the movement of the charity wards into private hospitals from county hospitals. This shift of charity wards has changed hospital-based patterns of care and made the private hospital a necessary component of the teaching array that the medical school uses. It has further required, I think, a rather reluctant movement away from the ivory tower.

More recently, the things that I think are reshaping the activities of the medical profession are the Professional Standards Review Organizations, the Health Maintenance Organization (HMO's), which are being started, and the certain occurrence of some kind of national health insurance. As you know also, beginning with the Academy of General Practice (founded some 25 or 30 years ago), efforts have been made by professional societies to upgrade the quality of the care by insisting on some continuing medical education as a requirement for membership. This is now much more wide spread, and most of the national specialty societies require certification on the part of the individual that he has participated in acceptable continuing medical education. You might ask "What is participation?" And I would agree with anything you would have to say about that. At this particular conference, for instance, the degree of participation of each of us here has varied from day to day and from hour to hour. The mere fact that each of us can certify that we have been at this conference for 4 days doesn't necessarily mean that we have participated intellectually or profitably in the proceedings. This exemplifies one of the really critical problems about the requirements of continuing medical education for membership in societies. Interestingly enough, there are two states which require a minimum participation of 150 hours per 3-year period of continuing medical education for relicensing periodically. These two are New Mexico and Maryland. I suspect there will be others.

We in the practice of medicine, at least the people I knew and worked with, tried to keep up without much success. We had books which were out of date when we got them and journals which we tried to read, although I recall mine constantly piled up in the corner of my office. That pile never decreased until I took a vacation and my nurse threw out the pile of journals so I could start again when I got back. And it wasn't because of lack of interest. The detail man is given credit for continuing the education of physicians, and I suppose this is true. My own feeling about detail men was that it was rather like a social call, and I preferred that he didn't talk about his products. If he wanted to come up and say hello, fine—I would see him at 5:00 or later.

The point I am leading up to is the fact that there are pressing reasons why we physicians must become more involved with those of you who are working so effectively in data management. But we have a real problem because we don't understand what you are doing.

I had reported to me at a recent conference held at the University of Washington that medical audits can be performed by computers for certain patient care categories. Total computer costs for that audit amount to $1.00 per case. On the other hand, an old-fashioned retrospective audit, involving 6 to 10 physicians sitting down and pouring over the completed patient charts, probably costs closer to $12.00 per case, if they are carefully done.

I wish that physicians and medical researchers could get together more often and exchange ideas and think together. I don't recall anyone from your field coming to me and saying: "Hey, let's sit down and talk."

However, I can remember when I was Director of Medical Education at the hospital in Eugene and I went over to the University of Oregon and said: "I've got 20 physicians over here that would like to sit down and talk about the chemistry of the atom." Amazingly, these physicians invested 3 hours of 1 night each week for a couple of months and were delighted with that they learned. Therefore, I don't think that talking with us is a totally hopeless task.

Programs to introduce medical students to communications technology and computers are also lacking. This problem is easily correctable if the attitude of medical schools would permit. I've asked some of the people here at this conference how they get medical students involved, and they have explained that medical students who are interested and motivated seek them out, and then as one computer researcher put it: "We try to suck them out of the main stream." This modus operandi does not constitute a general effort to acquaint practitioners with your research work, and we are the potential users of your findings.

I remember when I first came back to Bethesda 5 years ago, Dr. Davis asked if I would like to go out to a conference on communications being held at the University of Missouri. I spent a couple of evenings with Dr. Lindberg and Dr. Baruch, and I was speechless throughout the whole exchange between them. I was really quite frightened at what these guys were promising. (I'm still a little bit afraid of what they are now predicting.)

While at the Lister Hill Center, I was able to follow closely several interesting experimental applications of communication technology that to a degree involved computer technology. One of these, the

Alaskan Satellite experiment, utilized the National Aeronautics and Space Administration's (NASA) first Application Technology Satellite (ATS-1) which was used to relay messages between physicians at the Tanana Native Health Center Hospital and aides at remote villages in the Tanana District (Yukon Basin). For the first time in history reliable voice communication has been possible on a regular daily schedule.

On May 30, 1974, NASA put ATS-6 into synchronous orbit. Included in more than 20 experiments to be carried out via this satellite is a more sophisticated application of communication technology to health care in Alaska. Portable black and white TV cameras will be placed in villages and at the Tanana Hospital to add visual capability to the audio. ATS-1 will continue to function as part of the systems to access patient data contained in the computers at Health Programs System Center (IHS) in Tucson, Arizona.

The Lister Hill Center will also use NASA's ATS-6 satellite in a fully interactive color television mode to teach medical students in the University of Washington's WAMI program. WAMI is a regionalized medical school program at the University of Washington that admits 10 freshman medical students from each of the states of Alaska, Montana, Idaho and Washington each year. The students stay in their home state institutions for the first basic science portion of medical school; come to the University of Washington for their second and third years, and then go back to their clinical facilities at home for their final year. The purpose of the WAMI program is to encourage these students to later establish their practice in their home areas. The ATS-6 satellite enables medical schools to relay communications between these students and the faculty at the University of Washington, reducing travel time and expenses. (Not very welcome to the faculty who wanted to go to Alaska when the fishing was good.) Preadmission interviews of students and a variety of administrative functions will also occur. Computer assisted evaluation of student accomplishment is provided through interaction with the central data base at the University of Washington. Time does not permit a description of the other computer technology activities supported by the Lister Hill Center and the National Library of Medicine. (Medline and the Computer Assisted Instruction Network.)

My final thought is that the enlightened members of the medical profession, and there are many, have no fear about what you are doing. They would like your help because they are concerned at their own ever-growing data loads. Furthermore, I would like to propose that the next time we meet, half of the conferees be M.D.'s 35-years old or younger.

Progress in computer applications in the life sciences as seen by congressional staffs

ROBERT W. MAHER [1]
Rayburn Building
Washington, D.C. 20515

In reading the literature, I frequently come across authors alluding to the marriage between science and engineering in the area of our space program. A 20-year honeymoon is also often mentioned.

Then comes the obvious question — what will this happy union produce in the area of medicine and biomedical research?

I would like to be able to say specifically that *X* number of dollars will be forthcoming for certain programs. But in all honesty, I can only say that the situation is pregnant.

Technology in the form of the computer was advanced to serve as a tool in the space program. But without it, I am sure that mathematicians would still be figuring out problems on a stack of papers which would reach the moon.

But they had a single, identifiable problem — how to get to the moon.

In biomedical research and health services we have literally millions of problems involved — because we deal with the human factor — but without a single identifiable goal. And unlike the "cost be damned" attitude which accelerated the space program, we must operate under a cost-effective mandate which, as you well know, seems to get more and more restrictive.

We have, however, ample evidence that the computer is finding acceptability in many areas ranging from the very important breakthroughs in the area of biochemistry and the genetic code work, to simple but equally valuable record keeping in hospitals and information retrieval in the basic sciences.

Maybe some of the edge of expectation has been taken off by visions expressed many years ago that foresaw hospitals run entirely by computers. But to most, the potential of the computer is still great and exciting.

We find evidence in various legislation that the Committee on Public Health and Environment, chaired by Congressman Paul G. Rogers of Florida, has produced. The Health Manpower Training Act has provisions for the use of computers. The National Institutes of Health, of course, has a division of Computer Research and Technology and the National Center for Health Statistics and Library of Medicine both are heavily involved in more fully utilizing computers.

I hope that some day we will see the Institute for Computer Sciences and Technology and the Office of Technical Assessment take aim in the area of health and the life sciences.

---

[1] Administrative Assistant to Congressman Paul G. Rogers, D, Florida.

When I mentioned before the NIH and its present involvement, I think that great credit is due for foresight on the part of Dr. Shannon. Dr. Shannon ordered the first computer on faith, despite the fact that no one really had a firm idea of what it could be utilized for.

Well, we are well beyond the faith stage. For instance, the National Cancer Institute is organizing what must be the largest scientific data bank in a specific health area. Now, under the Office of International Affairs, the International Cancer Research Data Bank will come under the control of NCI in July when the present contract terminates.

And, of course, the National Heart and Lung Institute is employing computers for statistical data which will also help, I believe, the basic researcher.

I know that simple retrieval of information does not sound very exciting, but the potential impact on biomedical research and medicine in general can be great.

For instance, there is an on-going program called the Brain Information Center which disseminates information on neurological disease and disorders. Assuming that this is the repository of more knowledge on the subject than anywhere else, and assuming that there are significant inquiries, then the mere passing of definitive information to a practitioner or investigator does, I feel, advance the state of the art.

We see the use of the computer in the Regional Medical Programs for diagnostic treatment, and educational purposes. And this too advances the state of the art. For people who lacked it, the knowledge passed on the computer has expanded their capabilities.

I remember the first time I saw a computer application to medicine. A young man at the University of Miami had conceptualized a computer manikin and had it programmed to give all the symptoms of a heart attack. I was fascinated by the actual mechanics of the manikin and somewhat boggled by the potential. All the usual forms of heart disease that many physicians in training will never see in a patient during their education can be brought forth by the press of a button.

These are the real and practical tools which we now have available in the hospitals, clinics and laboratories.

But they are thinly spread and underfunded. Yet the potential is obvious. Unlike the great leap that took us to the moon from the visible concentration of technical resources in Cape Canaveral and Houston, we have only the campus at NIH to identify with. The space program was a coordinated national effort, but in health we are spread across the landscape and are progressing in a rather serendipitous fashion. We have no national plan, nor a national goal.

I would like to depart on what I think is a far more imposing problem in the area of life sciences than the application of computers and technology.

And that is the very basic concept of the value of biomedical research. As you know, the Administration proposed to terminate all research training grants and fellowships. Since there was no law on the books that actually established grants, it could do this. Each institute provided the funds that it deemed appropriate for its own training and fellowship programs.

So the Administration's proposal was not so much a rejection of something the Congress had produced as it was a rebuke of the worth of such grants. Testimony during our hearings did not at all substantiate the Administration's position. To the contrary.

The House passed legislation which Congressman Rogers had molded to put into law a mandated research training and fellowship program.

The long and the short of it is that in June 1974 we formally filed a report on this legislation. We have authorized \$208 million.

When it became evident that the Congress was going to pass this legislation, the Administration responded to Congressional ire by promising $30 million to finish out existing grants.

So we are going well beyond what the Administration wanted because the Congress believes that the research training program is the backbone of American science. And only through such a program can the United States maintain its preeminence in the area of biomedical research.

When the Administration challenged the program, the basic argument was that there was no proof that the training programs were working, and did in fact, draw bright young minds into research.

Despite the evidence presented during hearings, and a softening of the Administration position, we could envision another such assault on the program in the coming years. So the Congress conferees included in the bill a requirement to define the nation's needs for biomedical research personnel and the adequacy of existing training programs. The report is to be an annual one, to keep abreast of the existing situation.

During the past 6 years, those in the Congress who have had the responsibility of overseeing and guiding health programs have not had the luxury of looking forward and planning. They have continually had to look backward and expend time and energy in order to halt attempts on the part of the Administration to erode the progress that has been made.

The Administration and the Congress unfortunately have been cast in adversary roles. I do not mean to sound dramatic, but in the 8 years that I have worked with Congressman Rogers, I have seen him time and time again step in to fend off efforts on the part of the Office of Management and Budget or the Administration to cut back or terminate programs.

I wanted to mention that, because I now sense somewhat of a new thinking within HEW. This is due partly because of the massive setback HEW had in the area of impoundment last year. They lost 34 or 35 suits while holding up more than $2 billion in health funds.

Now I have a feeling that we may be headed for better times. I certainly hope so. For it is in the best interest of this nation and its people that we do move ahead for improvement of this nation's health — through sound and unencumbered basic research by developing an effective system that will pass on the knowledge gained in research to our health delivery system.

I guarantee that we will use every tool available to reach that goal. Right now, the computer stands as one of the most readily available tools that we have.

Closing remarks at FASEB conference on the computer as a research tool in the life sciences

DONALD LINDBERG
University of Missouri
Columbia, Missouri 65201

The variety and scope of the papers presented at this meeting do not permit a technical summary. There are, however, some general impressions which we might consider.

First there are the contrasts between this meeting and similar gatherings 10–15 years ago. The years have not brought a major change in the fundamental problems in biomedical computing. They have, on the other hand, brought a host of talented new investigators to the field, new software systems, and importantly very much improved computing equipment. Networks also now seem increasingly to be worthy of exploration by individual computer users and groups as a potential means to share computing resources. Network use has the additional advantage that a system can be maintained by a single group, thus bypassing the problems of transplantation of system and of updating and improving. These changes in the computing scene proffer potential new solutions to our old problems.

In contrast, we have heard in a number of papers and discussions of a new, even if not fundamental, set of problems: namely, how to live with success. Biomedical computing systems may or may not have changed the world significantly, but many of these systems have succeeded. That is, many have met or exceeded their design objectives. Now there are the new problems, often especially difficult in a scientific system: how to make a transition from research to practice? How to pay for successful operation when system function is not closely coupled to individual patient welfare? How to resist or react to the urgings of hardware experts continuously to tamper with a working system to keep it "current?" The Kaiser-Permanente data base system and the University of Alabama cardiac monitoring system seem especially striking examples of this dilemma.

Along with the acknowledgment of the problems of success comes an awareness that biomedical computing is now a valid and self-sufficient research area and that it absolutely is already a major tool in support of significant life science research. The work we have heard described is a substantial proof for this claim. The few days of our meeting permitted only a sampling of the contribution of computing to life science research, leaving much excellent work unmentioned.

The second set of general conclusions stems from the coming of age of biomedical computing as a research field. Even without a formal rite of passage, the field seems to have reached a stage at which certain responsibilities fall to us as investigators. What follows then is my own interpretation of some of these obligations.

We have a responsibility individually to seek out and explore solutions to our old research problems. What was insoluble 10 years ago may yield to the improved hardware of today—or perhaps to the new minds at work in the field now. What used to be an exotic service may now be a commercial time-sharing offering. Such offerings are burgeoning, and the means to keep informed are diminishing. The initiative remains with each of us individually.

We have a responsibility as biomedical computing people to keep a real sense of perspective as to proper computer applications, and honestly to keep telling our colleagues what is reasonable. Just two examples: we have heard several times during the meeting that successful systems have an almost equal dollar investment in people and hardware. This is no law of computing, of course; but perhaps a useful guideline. Another figure recurring in the meeting relates to the minimal budgetary cost for making any large data base system: a million dollars a year. If there are such features characteristic of computing research, then we must acknowledge them and plan accordingly.

Next are two responsibilities for computer people that are difficult to achieve but are traditional goals for any good scientist: to try to be humble and to try to be sophisticated. For us, to be humble means: to try to understand the principles and needs of biomedical computing as a field rather than only the specific needs in our own research; to remember that there is more than one way to do good computing—even more than one way to do a particular computing job well; and to retain a liberal viewpoint about budgets (even other people's budgets!) Some experiments can be done cheaply. Some computer applications, to be tried at all, are quite expensive. If the total dollars for research are really shrunken, it may be necessary that we agree to limit the number of places at which some expensive things are tried. An example may be large data base experiments.

Usually computer people have ample sophistication concerning the complexities of their executions. Professor Shipton reminded us to keep some sophistication also in our long-range planning: he threatened us that the trillion-bit memory may happen to us one day. Doubtless there will be intermediate surprising developments. His real warning was that we not plan on the basis of the limitations of the systems we are used to.

That we should make efforts to think in a politically sophisticated way was urged by Doctor Goldberg when he said that our working systems must be paid for from patient care fees and savings in health care costs. It was clear that whatever level of sophistication we were to achieve would need to be substantially greater than that reflected in current cost accounting schemes. Savings resulting from the use of proper computer systems and studies could easily accrue to the whole country, the county, or to a hospital over many years. Indeed it was clear that the most significant savings might not be reflected at all in a single patient billing or even an annual Blue Cross accounting.

One example from discussion at this meeting was the ultimate savings in life, deformity, and health care costs that stemmed from the record keeping systems concerning stilbestrol administration to pregnant women. The practice resulted in malignancies in the daughters 2 decades later. This knowledge will prevent further such errors. The cost of the record system increased the original patients' bills, but society as a whole gained from the record keeping and analytic function. The same will be doubly true for computer record system costs.

Dr. Baruch's talk likewise emphasized the need for political sophistication in our costing of computer developments and savings. Yet even in his analytic framework, the need to put a sufficiently long time constant over the costs and benefits would clearly become the critical issue.

One last example of a need for sophistication on our parts in the nontechnical area should be mentioned. This is our growing concern with the need for protection of personal privacy, especially in connection with the building of patient care computer files. I do not urge upon you any particular recommendation: that is, to build the files, or not to build them; to support legislation, or not to do so. Rather I urge that you each give the problem careful thought. It is possible that we shall have Federal law to guide computer file operations. Perhaps this will prove a helpful deterrent to misuse. But anything one person enciphers another can decipher. Any set of secret file names can be read. Truly secure installations could not use the methods most of us need: dial-up lines, shared files, automatic restart procedures. The actual shape computer installations take will almost certainly not be given by law but by the practices you and your colleagues adopt. Deciding on the final form—or being aware that a final form is reached—may cause us all some conflict. At the very least, we can be certain that there will be circumstances in which the needs for protection of privacy will demand that we ignore what we know to be the best computer or data processing techniques. We may need to argue against "collect data only once" and against "shared files" and "the unit record" and many practices that we have taken to be good and efficient. Another level of the coming testing of our loyalties may arise from the apparent legal ownership of certain files, even hospital files, by universities. Whatever form the problem takes, the real protection of society will ultimately reside in the qualities of the people who build and operate the data systems. You and your colleagues—not the courts—will be the bastions. If you will give the issue careful and professional thought, we can be confident you will find good and rational resolutions.

Observations on the Conference as seen by a British participant

D. E. CLARK
University of Manchester
Manchester M13 9PT, England

It was evident that there had been a considerable change in emphasis on the approach to computers as service and research tools in the life sciences.

There were those who had assumed that because computers in the less mathematically precise sciences of biology and medicine had not shown a direct and immediate return in the short term then it was of little use in pursuing their introduction further. Those who advocated this policy of short-term thinking would most probably be the first to complain that the latest methods of diagnosis and treatment were not being applied in their case when they came into contact with medical science.

Support for life science is a very long-term investment, a gradual change over a period most probably in excess of 10 years, i.e., change measured in decades not annually.

A quick injection of computing technology will have little effect on the life sciences. For example, a doctor has to learn to practice medicine with all but the simplest of machines and if they come and go in a transient fashion—he will not bother with them.

He has seen as many fashions occur in his own drug industry and the like. Long-term effort creates stability of approach in a subject.

Turning on and off of effort by a country merely wastes precious resources. A nation that sees the long-term advantage will gain significantly over the group who over-react. It is the old hare and tortoise problem.

The conference clearly showed the need to generate a new permanent force of biotechnology: a number of papers were able to illustrate the need in a computing project for: *a*) the life scientist, be he a Doctor of Medicine or a Doctor of Agriculture; *b*) computing specialist and scientist; and *c*) engineering expertise to help weld the project together. This team needs to be together for some minimum length of time.

The computer system that they designed to do the job, and was shown would do the job, would always do that job irrespective of the advances in computer technology.

Therefore it was not necessary to consider that the computer was a little out of date and should be changed. Computers were only one of the tools in the larger problem that biotechnology teams were looking at.

Specific areas in the conference where impact of new technology would be felt, would be in the interactive graphics system. National networks were shown to be fine for certain problems in the physical sciences but of little use because of long response time for many biological problems.

Both American and European groups noted the cost of large data base systems was very high; present figures did not allow any change from one million dollars per annum. Considered on the long term over a large population they saved patients expenditure and therefore were an economic gain to a nation. Given the right size computer in a life sciences program the cost per unit could fall as low as 0.2% when considered over a 10-year period.

Advocates of the minicomputer have often failed to realize that in areas of biological sciences, where this technology was continuously evolving, even a "service" must be looked at as long-term research.

Observers noted that a minicomputer user may, over a period of years, be trying to fit a quart into a pint pot, and should have started out with a medium-scale machine in the first instance—highlighting the problem of underkill and overkill.

Since biotechnology was a new science, then there was need for centers of high competence to assist the life scientist, including medicine, to a slow but steady shift over to a greater emphasis of measurement and analysis of problems, rather than purely the description of the problem.

The conference saw the need for mini- and microprocessors and a decentralization of computing that, because of difficulties of access, stifled growth.

There was a need to duplex machines in critical areas, for the experience of NASA had shown that such techniques were very necessary when man depended on machines for life support.

Biotechnology could and should have a strong impact on the need for more medical staff in the less attractive rural areas. This shortage was significant at present and would get worse as the demand for better care had to be met.

Computers in biotechnology across the board could not be said to be economic. Those who were in the field had many anecdotes to tell on the positive side but were unable as yet to come up with hard facts on dollars and cents saved.

Conference summary and conclusions

WILLIAM SILER
University of Alabama
Birmingham, Alabama 35294

AT LAST—A RELIABLE COMPREHENSIVE OVERVIEW OF COMPUTERS IN LIFE SCIENCE RESEARCH

The importance of this conference is immediately indicated by the prestige of its sponsors: the Federation of American Societies for Experimental Biology; the National Science Foundation; and the National Institutes of Health. Although there have been many conferences and publications on the use of computers in the Life Sciences, most notably the series of volumes on *Computers in Biomedical Research* (1), this is the first time that a conference has been undertaken which spans all aspects of the field from basic science through clinical medicine, with considerable attention to the costs of using computers, and with viewpoints ranging through those of life scientists, industry, scientific administrators, and the Congress. For the first time, a set of papers has been collected of sufficient breadth on which to begin to base a national policy regarding the use of computers in life science research.

This summary, based not only on the presentations themselves but also on innumerable discussions with the participants (some of which continued for some weeks after the conference), will attempt to distill the consensus of the participants, including those not expert in the use of computers, and those heavily concerned with the costs and benefits of their use. Of course, not all the participants were in agreement on everything; nevertheless, by the close of the conference, a surprising degree of unanimity had emerged.

UTILITY OF THE COMPUTER IN LIFE SCIENCE RESEARCH

Almost 10 years ago, although computers had already begun to have a significant impact on biomedical research, it was stated (2) that "There have been no major biomedical breakthroughs directly attributable to the computer." It is apparent from the papers at this conference that this is no longer true. At opposite ends of the research spectrum, we have the major breakthroughs in the understanding of the structure and function of biological macromolecules, in which the computer was an absolutely essential element; and in massive clinical practice employing computer-based multiphasic health screening, making it possible to plan totally new health-care delivery systems using the computer greatly to increase physician effectiveness. In the laboratory, the minicomputer has already become an instrument considered by many to be as essential as a good preamplifier. The availability of large-scale sophisticated statistical packages has totally changed the way in which the biostatistician works. The demonstrable benefits of the use of computers in life science research over the last 10 years can leave no doubt that the computer is achieving acceptance in life science research, not as a status symbol, but as an essential tool.

However, the results achieved by the use of the computer have not been achieved without considerable effort on the part of a good many people, and also at considerable cost. We now see under way strong efforts in computer science at rendering computers more cost effective, especially through the use of networking; and at the theoretical end, the efforts in the field of artificial intelligence at knowledge organization and manipulation so that the complex problems of biology may begin to be expressed more directly in terms of biological knowledge, and still be acceptable to that "supersonic moron," the computer.

It seems, then, apparent that a new field now exists, that of the fruitful and cost-effective use of the computer in life science research. However, as one of the speakers pointed out, this field seems so large and complex as to defy any one person's understanding. It is apparent from the proceedings of this conference that a great deal is now known about this field, but that this knowledge is dispersed among a number of talented individuals in a somewhat haphazard way. It would seem eminently worthwhile to determine whether or not this field can be organized in such a way as to permit training those scientists (and perhaps administrators) who are likely to be involved in computers in life science research.

Let us examine the kinds of persons who have proven efficient in the life science use of computers. First, we have the person who was fundamentally trained as a life scientist, who possesses a certain amount of mathematical aptitude, and has gone about achieving computer competence in his own right. Such "one-man-bands" have proven probably maximally efficient, not only in their own work with computers, but in their ability to communicate with mathematicians and computer scientists, and hence not infrequently to lead effective research teams. Secondly, we have the cross-disciplinary professional, trained in mathematics, computers, and the basics of the life sciences, able to communicate with biomedical researchers on their own terms, translate their problems meaningfully into computer terms, and then to solve them. Depending on orientation and institution, such professionals may be called biomathematicians, mathematical biologists, biostatisticians, theoretical biologists; the generic term "biomathematician" will be used here. Last, we have the computer scientist who interacts actively with members of the life science community, and is responsive to their needs in the production of hardware and (to a lesser extent) software suitable for their needs.

In the hardware development field, the life scientist also needs cross-disciplinary support. While the networking effort can be carried on by computer scientists, physically interfacing the minicomputer to experiments is optimally done by specialists trained in life science basics, electronics and computers—the new bioengineers. Even then, the conventional minicomputer may be far from the optimal laboratory computer node; two of the papers at this conference are concerned with the special design and configuration of laboratory computers, both efforts being spearheaded by dual-trained life/computer scientists.

In short, if the success of computers in the life sciences is to become a science rather than an art form, future training of life scientists should include training in biomathematics and biostatistics, and in computer techniques; and most badly needed of all (partly because they will be competent to train life scientists in quantitative techniques) is the training of biomathematicians knowledgeable in the basic principles of the life sciences, deterministic and stochastic mathematical techniques, and in computer methods.

In both cases, thorough grounding in the principles of inductive inference as a scientific tool is important in understanding the problems of extracting new knowledge from noisy data.

There are, unfortunately, several obstacles to mounting such training programs. There is the crippling of training grants in recent years; the inertia of the classical life scientist who has got along quite well without computers, thank you, and who feels no strong pressure to adopt new methods; the inertia of existing curricula; and the hard fact that to demonstrate locally the power of computer techniques it takes not a little, but a lot of special training, with no guarantee of success even then.

It is impossible to assess quantitatively how much the lack of such training is costing the nation in terms of missed discoveries, inadequately interpreted data, and wasted time in experimentation. It is the experience of this author and other biomathematicians whom he has contacted that no more than about one-quarter of the problems in data analysis initially presented as straightforward statistical problems actually yield their full value to standard statistical techniques. The ability of a properly-built model to extract meaning from otherwise baffling data, and considerably to increase efficiency of experimentation, has been demonstrated over and over again.

If the computer is to be used effectively in biological research on a mass scale, and if the time spent by skilled biologists in experimentation is to be most fruitfully used, a very strong national effort should be mounted towards the twin goals of training life scientists in quantitative and computer techniques, and in training expert biomathematicians who can successfully interface the computer with life scientists who lack thorough quantitative training. The importance of this goal cannot be overestimated.

Further, the scarcity of research dollars dictates that training in the cost-effective use of computers should be included. This means training in the available options in a dynamic way, since the cost picture is changing rapidly. Our new scientists must be able to determine with some confidence whether a job should or should not be done on a computer, and if so, what kind of a computer should be employed, and what mathematical techniques will yield maximal new biological knowledge. Their success in obtaining research funds will depend in part on whether or not they are using computers in a cost-effective manner.

CONCLUSIONS

1) In skilled hands, the computer can be of great benefit in life science research, with extremely important computer-dependent work being done in fields ranging from molecular biology to the practice of clinical medicine.

2) Computers can no longer be justified as status symbols; they must be used in a cost-effective manner.

3) The most serious handicap to the widespread effective use of computers in the life sciences lies in the lack of supply of trained personnel (call them biomathematicians) who are expert in both deterministic and stochastic mathematics, computer techniques, and the fundamental principles of the life sciences, who can be successful in interfacing currently-trained life scientists to the power of the computer; and in the lack of training of life scientists in mathematical and computer techniques.

4) Also needed are bioengineers and computer scientists with cross-disciplinary training to design, construct and interface all the computing and electronic equipment necessary for cost-effective research, and theoretical computer scientists to develop new and more effective methods of problem statement and solution.

5) Given a supply of scientists trained in this cross-disciplinary fashion, the computer should be able greatly to increase the efficiency and creativity of life science researchers.

6) The individual computer-life-scientists have done their job by demonstrating the remarkable power of the computer and associated mathematical techniques in a wide variety of fields; it is now up to the Congress, granting agencies, deans and department chairmen to implement the broad training and research programs necessary to make this power widely available and useful in life science research.

REFERENCES

1. Stacy, R. and B. D. Waxman (editors). *Computers in Biomedical Research*, Vol. I through IV. New York: Academic.
2. Waxman, B. D. Biomedical Computing—1965. *Ann. N. Y. Acad. Sci.* 128: 723–730, 1966.

Index

U.C.W. ABERYSTWYTH LIBRARY